W9-ATT-620

PEDAGOGY, RELIGION, AND PRACTICE

Pedagogy, Religion, and Practice

Reflections on Ethics and Teaching

Alan A. Block

PEDAGOGY, RELIGION, AND PRACTICE: REFLECTIONS ON ETHICS AND TEACHING
Copyright © Katharine Adeney, 2007.
All rights reserved. No part of this book may be used or reproduced in any manner
whatsoever without written permission except in the case of brief quotations
embodied in critical articles or reviews.

First published in 2007 by
PALGRAVE MACMILLAN™
175 Fifth Avenue, New York, N.Y. 10010 and
Houndmills, Basingstoke, Hampshire, England RG21 6XS.
Companies and representatives throughout the world.

PALGRAVE MACMILLAN is the global academic imprint of the Palgrave
Macmillan division of St. Martin's Press, LLC and of Palgrave Macmillan Ltd.
Macmillan® is a registered trademark in the United States, United Kingdom and
other countries. Palgrave is a registered trademark in the European Union and
other countries.

ISBN-13: 978-1-4039-8373-2

Library of Congress Cataloging-in-Publication Data

Block, Alan A., 1947-
    Pedagogy, religion and practice : reflections on ethics and teaching /
Alan A. Block.
        p. cm.
    ISBN 1-4039-8373-9 (alk. paper)
    1. Teaching. 2. Reflection (Philosophy) 3. College
teachers--Professional ethics. 4. Education--Philosophy. I. Title.
    LB1025.3.B587 2007
    371.102--dc22
                                        2007003545

A catalogue record of the book is available from the British Library.

Design by Scribe Inc.

First edition: September 2007

10 9 8 7 6 5 4 3 2 1

Printed in the United States of America.
Transferred to Digital Printing in 2010

Dedicated to my dear friend, Gary Welch, who,
over hundreds and hundreds of miles has
listened to and discussed every word of this book.
I thank him always with insufficient gratitude.

CONTENTS

Adonai spoke to Moses on Mount Sinai, saying: Speak to the children of Israel and say to them: When you come into the land that I give you, the land shall observe a Sabbath rest for Adonai. For six years you may sow your field and for six years you may prune your vineyard; and you may gather in its crop. But the seventh shall be a complete rest for the land, a Sabbath for Adonai; your field you shall not sow and your vineyard you shall not prune. The aftergrowth of your harvest you shall not reap and the grapes you had set aside for yourself you shall not pick; it shall be a year of rest for the land. The Sabbath produce of the land shall be yours to eat, for you, for your slave, and for your maidservant; and for your laborer and for your resident who dwell with you. And for your animal and for the beast that is in your land shall all its crop be to eat.

Leviticus 25:1–7

Walden is melting apace. There is a canal two rods wide along the northerly and westerly sides, and wider still at the east end. A great field of ice has cracked off from the main body. I hear a song-sparrow singing from the bushes on the shore—*olit, olit, olit—chip, chip, chip, che char—che wiss, wiss, wiss.* He too is helping to crack it. How handsome the great sweeping curves in the edge of the ice, answering somewhat to those of the shore, but more regular! It is unusually hard, owing to the recent severe but transient cold, and all watered or waved like a palace floor. But the wind slides eastward over its opaque surface in vain, till it reaches the living surface beyond. It is glorious to behold this ribbon of water sparkling in the sun, the bare face of the pond full of glee and youth, as if it spoke the joy of the fishes within it, and of the sands on its shore—a silvery sheen as from the scales of a *leuciscus,* as it were all one active fish. Such is the contrast between winter and spring. Walden was dead and is alive again. But this spring it broke up more steadily, as I have said.

Henry David Thoreau

And how can I not call it "Melting Apace." Perhaps this process refers to me as well—as the sabbatical proceeds, perhaps I, too, will melt apace and the life in me will reawaken.

Alan A. Block

INTRODUCTION

This book is a work of teshuvah, a Hebrew word the meaning of which remains yet open to me. All of my life I have understood the word teshuvah to mean "repentance." Repentance, as we well know, is sorrow for our misdeeds, our misthoughts, our mistakes. Martin Buber, however, suggests that teshuvah might be understood as something other than "repentance." Teshuvah can be understood as the name given to the act of turning to the world to work towards its redemption. Teshuvah, according to Buber, is more than an act of personal contrition; it is, rather, an act of personal commitment. In teshuvah, we turn toward the world to begin the healing that our ethical absence from it has made necessary. In the process of *tikkun olam*—the repair of the world—we achieve as well, our own healing. "Teshuvah," Buber says, "is the name given to the act of decision in its ultimate intensification; it denotes the decisive turning point in a man's [sic] life, the renewing, total reversal in the midst of the normal course of his [sic] existence"(1967, 67). Teshuvah is the decision to change one's life. The *Pirke Avot* says, "One hour of return in this world is better than the entire world to come"(2002, 4:22). This book, *Pedagogy, Religion, and Practice: Reflections on Ethics and Teaching*, is an account of that hour.

This book is a work of teshuvah. There are precedents for such redemptive acts. Henry David Thoreau retired to Walden Pond to effect a return to his life. On Walden's shores, Thoreau meant to discover of what his life consisted so that no matter how mean it might be, he could yet grapple it to his soul. "There is more day to dawn," he cried triumphantly. "The sun is only a morning star" (1970, 448). *Walden* is the product of Thoreau's sabbatical and his work of teshuvah. When his father died, Leon Wieseltier used the year of mourning to effect his spiritual turning. At the end of that year, he wrote, "I looked above me, I looked below me, I looked around me. With my own eyes, I saw magnificence" (1998, 585). Wieseltier's *Kaddish* testifies to his wonderful exercise of teshuvah.

In the spring of 1999, I embarked on the sabbatical leave that I had been awarded at the University of Wisconsin–Stout. I descended into my basement to meditate upon my life as a scholar, as an educator, as a father, and as a Jew. This work is the account of my return. This work is a recording of what I did and thought during that sabbatical leave. This work is a reflection on my role at the university when I was not actively involved in it, on my role as an educator while I absented myself from the classroom, and on my role as a parent as my presence in the household and in our children's daily lives increased dramatically in an interestingly transformed manner. This work is an account of what I did during my sabbatical and what I thought about my life during that year. Rabbi Nethunia the son of Ha-Kanah said, "Whoso receives upon himself the Yoke of the Torah, from him the yoke of the kingdom and the yoke of worldly care will be removed; but whoso breaks off from him the yoke of Torah, upon him will be laid the yoke of the kingdom and the yoke of worldly care." (*Pirke Avot*, 2002, 3:6). Does Rabbi Nethunia assert that the burden of study frees one from the world, or that the burden of the world ceases to oppress those who accept the yoke of study? Are my responsibilities to my children the assumption of the yoke of study or my abandonment of it? Is the classroom the place where I meet the world and abandon study, or is the classroom where the engagement in study promises the world? What is the relationship between my study and my work? What is the purpose of study? What are the obligations of study and of work? What should education do for our children—for ourselves? What should the teacher do? Who (or is that how?) should the teacher be? When I threw off my worldly care and began my sabbatical, was the yoke of the kingdom and worldly care also thrown off? What could the act of study and scholarship mean to we who live in this world?

Rabbi Tarfon was wont to say, "It is not for thee to complete the work, but neither art thou free to desist from it . . ." (*Pirke Avot* 2002, 2:20). This work is a beginning that will not be completed. It is the work of teshuvah. During these months, I read and I studied texts central to Rabbinic Judaism, texts foreign to, or at least unacknowledged by, those engaged in the practice of Western education. Under the stimulus of these texts, I interrogated my life in order to change direction in the midst of it. To the extent that I have succeeded, I see everything from a different direction. To the extent that I have succeeded, I work now in a different place.

In the first part of the book—*Leaving the Fields*—I have written of my grasping and grappling with issues concerning the sabbatical leave, with the issues it raised regarding the relationship between pure scholarship and the more messy work of the world and with issues that arose concerning my changed existence in the household now that I was home all of the time celebrating my sabbatical. The second and much larger part of the book—*On Entering the Fields*—uses a long piece of the Babylonian Talmud, particularly, *Bava Metzia* (83B–87B), to structure my thinking about just such issues. At times I deal directly with the particular section of Talmud; at other times I use the particular *sugya*[1] to elaborate other issues that troubled me about my life of scholarship, of school, of family, and of faith. I have presented the entire Talmud section in its exact arrangement, though as you will see, the connections between sections are often obscure. I have often talked about these discrepancies in the text.

When Thoreau left Walden it was because he had other lives to live. Upon the completion of Wieseltier's year of mourning, he stood with awe and wonder in the world and experienced blessedness. Now I, too, have finished my sabbatical. I think, I realize my turning.

BEGINNINGS AND THE LAW OF THE TEN SAPLINGS

When does the sabbatical begin? Is it on the first day of leave, or is it some time when the reality of the sabbatical settles on the consciousness? These are certainly two moments related but not coincident. The former occurred on the twenty-first of January, 1999, when I descended into the basement office I maintain in my home. The latter event is one of the subjects of this book. When I came down here in January, it was with a firm but vague purpose. I suspect that preparation for sabbatical is meant to ensure the fullness of the leave—reservations to make, maps to study, appointments to establish, and books to purchase. But I often wonder about such complete preparations. The world is too full of mishaps. To plan too extensively is to factor out contingency and surprise. In the classrooms, where the emphasis is placed on the imposition of standards and national curriculums, contingency is factored out; this is too bad. Contingency offers us the possibility of another life! Before the sabbatical began, I planned carefully but not too extensively. And during the celebration of the sabbatical, I sat with my books and journals and fountain pens and inks. I read and I wrote. But I soon realized that I was not producing the book I had planned. The morning mists hovered about my feet. I was nervous. I continued to write. And at some point I understood what I was writing. It was high time to begin the sabbatical. But when was that exactly?

When does a Sabbatical begin? The Sabbatical year is when the fields are to observe a Sabbath rest. The sabbatical is when I must leave the fields. This issue was not unimportant to the Rabbis of the Talmud, because they were attempting to understand the practice of the Sabbatical mandated in the Torah. The Torah (Leviticus 19) commands that the Israelites be a holy people; their daily practices must aspire to that goal, though as the Prophets suggest, there was always great gaps between the ideal and the reality. I think the Rabbis were attempting to discover—or to invent—how to fulfill the commandments that would lead to holiness, and so they were intimately involved in the details that comprise the practices of daily life. The Sabbatical was not a small disruption of that mundane existence. Indeed, the sabbatical would mandate a dramatic change in the nature of daily life; it would require a complete refocusing of existence, as the practices of daily life had to be wholly reconceived. But if the commandment to celebrate the Sabbatical was to be fulfilled, then there were issues of daily life that first had to be resolved. And one of the clearest issues that had to be dealt with concerned the date for the beginning of the sabbatical

If the Levitical command demands a Sabbath rest for the land, then when exactly must we cease toiling the land? The Sabbatical year has to begin sometime, we know, but its exact commencement had to be formally established. There would, after all, be ways to alter the nature of sabbatical if one prepared carefully in advance for it. For example, there would be opportunities to circumvent the sabbatical commandments and turn a huge profit from the fields if the date for the beginning of the sabbatical was simply declared the first day of the year. A clever and prosperous agriculture worker could plow and sow enough grain immediately prior to the sabbatical year to make a considerable return from the growth in the eighth year as the fields worked (alone) during the seventh, though the worker enjoyed his rest. Thus, the Rabbis reasoned, if the Sabbatical year was meant as a Sabbath for the land, then the last plantings and plowings had to occur well before the new year formally began so as to ensure that the land received its ordained rest. When do the fields begin the Sabbath rest for the Lord? When must all preparations of the fields cease? On what day should the sabbatical year begin? The two dates are related but not necessarily coincident.

The Rabbis declared that the Sabbatical year always begins at Rosh Hashanah, the first day of Tishrei, the seventh month of the Jewish calendar. They derive this date by interpreting two Biblical verses, a common Talmudic hermeneutic principle. The first verse, in Numbers 29:1, declares, "The first day of the seventh month shall be a holy day to you . . . a day to you of blowing the Shofar." But, the Rabbis asked, what is the nature of this holiday declared in Numbers; what is the holiday's theme, as it were? To clarify this essence, the Rabbis looked in *Tehillim*, the book of Psalms. Psalm 81 states that we "Blow the shofar at the moon's renewal, at the time appointed for our festive day." Thus, since the only holiday that occurs at the new moon *and* on which the shofar is blown is Rosh Hashanah, the Rabbis declared that it was Rosh Hashanah of which the Psalmist spoke.

But though it is declared a holy and a festive day, of what does its holiness consist? What is so special about Rosh Hashanah? And furthermore, how do we derive the beginning of the sabbatical year from Rosh Hashanah? The Rabbis respond that Rosh Hashanah is a day of judgment. Psalm 81:5, quoted above, says, "The festive day . . . is a decree for Israel, a judgment [day] for the God of Jacob." Thus, the first of Tishrei, Rosh Hashanah, is the day on which Israelites are to be judged. Since on this festival day there is a judgment for the coming year, then this holiday is a *new* year's celebration. Furthermore, Deuteronomy (*Devarim*) 11:12 declares of the land to which the Israelites will go, "The eyes of God . . . are always upon it, from the beginning of the year to year's end." Thus, the Rabbis interpret, Torah declares that God's judgment begins at the new year, the festival that we know to be Rosh Hashanah. Finally, the Rabbis remark that the word in this verse for year, *shanah*, identifies the new year with the first of Tishrei; it is the same word *shanah* that identifies the sabbatical year. But the seventh year (*shanah*) shall be a complete rest for the land. Since the same Hebrew word, *shanah*, is used to refer to the beginning of the year of judgment and to the year of the sabbatical, the Rabbis reason from this repetition that the beginning of the Sabbatical year is to identified as the first of Tishrei, on Rosh Hashanah.

Which is all well and good for when the plow and pruning hook is to be formally laid away. But, as I have suggested, a very clever farmer could use the time wisely and plant right up until the beginning of the Sabbatical year and thus avoid paying tithes and taxes and still produce a huge crop, albeit, I acknowledge, with a few more weeds in his garden. To prevent this eventuality and to ensure that the land celebrates a true Sabbath, the rabbis stated that work on the land is forbidden not only during the sabbatical year but during the final days of the sixth year as well. Indeed, the Rabbis decree that all work must cease in a grain field after Pesach (which is celebrated almost six months earlier than the formal beginning of the sabbatical year), because until Pesach most of the moisture from the winter rains remains in the field, and plowing the still-moist fields would clearly be undertaken for producing growth during the sixth year. But work in the fields after Pesach, after the moisture has all left the soil, is clearly intended to reap produce during the Sabbatical year, and this is expressly forbidden; in the seventh year the fields are to be given a Sabbath rest. Similarly, work in the orchards is forbidden after Shavuoth, (four to five months before the Sabbatical year), an orchard defined as three trees equally distributed within a *bait se'ah* (2,500 square cubits or 600 square meters) and that can produce a minimum crop, as for example, enough figs to make a fig loaf weighing approximately 480 grams: a one-pound loaf of bread. Basically, then, the Sabbatical forbids use of the land beyond what might be considered its natural functioning. The land is to be left to rest, and the Rabbis in their rulings ensure that no one circumvent the Sabbatical rest by ingenious late plowings and plantings.

Except . . . the Rabbis state that along with the laws of the sabbatical year, the law of the ten saplings was also given to Moses on Sinai and authorizes some exception to the sabbatical strictures. The law of the ten saplings states that a special allowance for plowing may be made in the case of saplings, which must not be allowed to wither in their infancy. The law declares that if ten saplings are dispersed *randomly* in an orchard of 2,500 square cubits, or an area of 600 square meters, then the entire orchard may be plowed until the day before Rosh Hashanah, the actual nominal beginning of the sabbatical year when, of course, all work must formally cease. The explanation offered for the law is that without the plowing after Shavouth and up to the commencement of the Sabbatical year when all formal plowing in the fields is forbidden, the saplings would wither and die. For the sake of these saplings (the youth, as it were), work in the orchard is permitted. The law employs a structure (an orchard) to define non-structure. That is, the saplings must be dispersed throughout the field; if they are planted close together, say, in a circle, then they are not *dispersed,* as it were, and then the field may not be plowed.) This is done in order to enable work that nurtures.

The law is like this book: a structure that defines non-structure. Ten saplings dispersed make an orchard. Ten chapters dispersed make a book. When does a book begin? When did the sabbatical begin? Only that produce that does not violate the sabbatical commands may be eaten. Only that fruit is kosher.

Ten saplings. An interesting number. Ten commandments. Ten plagues. Ten garments that God is believed to wear, representing ten modes of behavior God employs for the salvation of Israel. Ten righteous people discovered in Sodom to save the city. Ten to make a minyana community of pray-ers. Ten chapters?

When does a sabbatical begin? By the law of the Ten Saplings, the Rabbis declare that the beginning depends somewhat on the nature and produce of the fields. Due to the nature of my fields, I am working right up until Rosh Hashanah. I have ten saplings in my field, and they require tending. I have continued to plow my fields for their sake. A great deal has been turned over. And these ten chapters dispersed throughout the field that is my life have sanctioned the plowing I have done. They are dispersed through the orchard; if you discover a pattern, then it is yours.

"I am on sabbatical. What am I to do?" I asked. What follows is what I did.

NOTE

1. A *sugya* is a sustained unit of deliberation.

PART I

LEAVING THE FIELDS

CHAPTER 1

THE FIRST SAPLING

I am on sabbatical leave from the University of Wisconsin–Stout for the spring semester, 1999. People who do not know me, or who are merely casual acquaintances and whom I meet casually and infrequently, or people whose memories fail them and who have forgotten that I am on sabbatical, enter casually into conversation inquiring, "So, what is going on in the office?" Or they ask, "How's work?" I respond that I do not know, that I am *"on sabbatical."* Inevitably the next question is proferred: "So, what are you doing?" I know that this type of question, like asking about the weather, is an effort at civility and is offered as an attempt to make some contact; it is an entry into conversation. But I have been continually taken aback by the question and confronted with the perplexities that beset me regarding the substance of this sabbatical. I often wonder myself, what exactly am I doing? Indeed, I have a research plan; one cannot be awarded a sabbatical at Stout without a formal detailed agenda. And I am to an extent following the research plan: I am reading in Jewish texts and I am trying to make connection to my educational practice. I had originally set out to integrate Jewish discourses into curriculum conversations—I had actually begun that work several years ago with a paper entitled, "Finding Lost Articles: The Return of Curriculum" (Block, 1997). That pursuit has continued with more and less success over the past years. It would have, indeed, continued *without* the sabbatical leave. My research is never very far from living my life. How could it be?

But I soon understood that to explicitly follow the developed plan would not accord with the motives behind the celebration of the land's sabbatical. Rather, this planned endeavor would mark a transference of the terms of contract, but would not alter the reality of the contract itself. And the sabbatical described in Leviticus 25:1–7 abandons all business compacts.

> Adonai spoke to Moses on Mount Sinai, saying: Speak to the children of Israel and say to them: When you come into the land that I give you, the land shall observe a Sabbath rest for Adonai. For six years you may sow your field and for six years you may prune your vineyard; and you may gather in its crop. But the seventh shall be a complete rest for the land, a Sabbath for Adonai; your field you shall not sow and your vineyard you shall not prune. The aftergrowth of your harvest you shall not reap and the grapes you had set aside for yourself you shall not pick; it shall be a year of rest for the land. The Sabbath produce of the land shall be yours to eat, for you, for your slave, and for your maidservant; and for your laborer and for your resident who dwell with you. And for your animal and for the beast that is in your land shall all its crop be to eat.

A complete rest for the land. A *complete* rest. Even the aftergrowth of my harvest is forbidden to be used for business or profit; rather, what grows in the fields during the sabbatical must be available for all to eat after the second rains. Even the grapes I had set aside for myself I may not pick for my gain. *All* labor on the land is forbidden. The sabbatical is indeed, meant to be a celebration.

✡

I am on sabbatical leave from the University of Wisconsin–Stout for the spring semester, 1999. Every day for the past thirty years I have gone to school. For thirty years, my first thoughts were of the classroom and the students who were compelled to attend it. For thirty years, I have lived often in a narrow place. Why is this semester different from all other semesters? Once we were slaves and now we are free. No sooner did the Israelites leave Egypt than they wished to return to the suffering with which they were at least familiar. They complained; they were a stiff-necked people. They would not make the effort to attain the land of the Canaan, the land promised to them. During this sabbatical, I think I can think about myself for a time and try to attain some promise of renewal.

I am on sabbatical leave from the University of Wisconsin–Stout for the Spring semester, 1999. There are things I have long longed to recon-sider. "But if the Blessed Holy One had not brought our ancestors out of Mitzrayim,[1] we and our children and our children's children would still be enslaved to Pharaoh in Mitzrayim" (Levy, 1989, 28). I read that the first sabbatical occurred at Harvard in 1880 when Charles Eliot, the President of the University, offered the promise of a sabbatical year to philologist Charles Lanham in order to entice him to come to Harvard (Zahorski, 1994, 6). I am glad that academia has institutionalized the sabbatical; our privileges seem always to derive from the efforts of others who preceded us. The Hebrew word for Egypt is "Mitzrayim" and it means "the narrow place." For this brief time of the sabbatical, I might consider that I have gone out from Mitzrayim. Perhaps I can use this sabbatical to consider the narrow places of the classroom. Perhaps, I might consider my own narrow places.

I am on sabbatical leave from the University of Wisconsin–Stout for the Spring semester, 1999. In the Fall, I will return to the classroom. And when I return to the classroom, what will I say of the time spent during this exodus? "So even if all of us were wise, all of us understanding, all of us old and venerable, all of us learned in Torah, it would still be a mitzvah for us to tell the story of the Going-Out of Mitzrayim. And everyone who enlarges on the Tale of the Going-out of Mitzrayim deserves praise." I can use this sabbatical to re-focus—focus again—on teaching and learning. When I return in the Fall, perhaps I may use my past to create my future. From the second night of Passover, celebrating the Exodus, until the holiday of Shavuot, celebrating the Revelation at Sinai, observant Jews count what is known as the *omer*. It is a measurement of barley. I read in the *Hillel Haggadah*:

> On the second day out of Egypt, our ancestors took a small measure, an *omer*, of barley, representing perhaps the small degree of holiness they had acquired since emerging from slavery the day before. From then until [the Revelation] they counted the days [with the measure of barley—the omer.]

Some say that every day they looked back on the progress they had made since that first full day and that first small measure; others say they relished each day for its own sake, its own growth, as days grow into a week and one week grew into seven.

> How have we grown, what have we learned. . .? As we begin to count the days till Revelation, let us begin to celebrate our small measures of religious progress as well (Levy, 1989, 123).

I, too, would measure my progress during this sabbatical, but what is it that I would measure? And what shall I use as my standard?

NOTE

1. *Mitzrayim* is the Hebrew word for Egypt. As I will explain later, *mitzrayim* means "a narrow place."

CHAPTER 2

THE SECOND SAPLING

THEM BONES, THEM BONES, THEM DRY BONES

We arrive home from the Purim celebration late, a bit inebriated from alcohol and chocolate and redemption. Amidst the levity bordering on chaos, we read the Megillah Esther on Purim, a story steeped in controversy and interpretation. For me, *Megillot Esther* is a narrative about the courage displayed by Queen Esther imperiling her own life to save the Jewish people. Purim is a holiday that celebrates how the Jews of Shushan were saved from extermination by the courage and cleverness of Queen Esther. There is much to revel about and much to observe. We arrive home exhausted. But Emma's homework remains incomplete.

What homework would that be? Her class is studying the human skeleton. In their classroom hangs "Bones," the school's full-scale model of same. Each child had been given an individual paper exemplar of "Bones," which they were to cut out for assemblage in school. Tibias and fibulas are relatively easy, but cutting out between the ribs is time consuming and tedious and requires very fine motor skills, not to mention very sharp scissors. Before leaving for shul last evening, Emma had cut out several lower ribs, but there was yet the upper rib cage and the entire sacroiliac and attendant bones still to release. I had been instructed to awaken her early (6:30 a.m.) so that she might finish her homework, but as a result of our celebration of Purim, I chose to let her sleep late. I considered: *there is little my daughter needs to learn from cutting out these dry*

bones and so I will complete the assignment for her. Perhaps a homework
assignment that can be accomplished by the adult without leaving some
trace of adult expertise is either an inappropriate or a meaningless task.

And so, while the oatmeal cooked and the toast burned, I stood at the
kitchen counter with a very sharp, single-edged razor blade and set free
the skeleton from its paper cage. I stared at the scattered pieces on the
wooden cutting board. I thought of the lines in Ezekiel 37:7, "Then
behold, and there was a noise, and the bones drew near, each bone to its
matching bone." But alas, they lay there on the cutting board immobile
and strewn in disarray. There sprang no skeleton into which spirit might
be blown. I stared still at the counter and again, I heard the prophet
Ezekiel: "Son of Man, these bones—they are the whole House of Israel.
Behold, they are saying, 'Our bones are dried out and our hope is lost;
we are doomed.'" I certainly hope this is not so; perhaps the spirit to be
breathed into these bones awaits my daughter's bones in school. As did
the bones in Ezekiel's vision, these bones, too, I believed, would come
together again, albeit in Emma's classroom. I worked meticulously frac-
turing only slightly the right clavicle.

I read in Kenneth Zahorski's book *The Sabbatical Mentor* (1994) that
one benefit of sabbatical leave is that it provides time for reflection. I
think I know what that means—the absence of daily university activities
leaves me free for other activities, one of which is reflection. To think! I
wonder what Mr. Zahorski thinks I have been doing all of these years! I
assume Zahorski refers to the activity of unbounded time free from the
daily activities of the university—classes to prepare, meetings to attend,
papers to grade, meetings to attend, students to counsel, meetings to
attend, research interests to pursue, meetings to attend, papers to write,
and meetings to attend. But I am not certain how one goes about *doing
reflection* during this sabbatical leave. I am not intending to be specious—
I mean, I have read the life of Thomas Merton. I have friends who regularly practice Transcendental Meditation; I have myself been running
for my life on the roads over the past thirty-two years. I regularly replenish my supply of endorphins. Does Zahorski mean that I will think about
things in my freedom that I did not—and could not do—in my mundane existence? But if so, how could I possibly plan for this reflection?
Had I written in my application for sabbatical leave that I intended to
sit quietly and await new thoughts, I doubt that I would have been
awarded the leave. Or does Zahorski mean that I will have now sufficient
time to engage in extended periods of solitary contemplation unencum-
bered by mundane issues? But in the non-monastic life I choose to lead,
quiet and repair from the world occur in relatively brief moments. For

example, in just a short while I will go upstairs to make lunches for my two children. And then we must discuss energetically, and with some friction, what clothes to wear. I mean, the world goes on up there. Thank goodness. Sometimes my serious reflection is disturbed by the mundane. Thank goodness!

And besides, the production of a thought occurs sometimes in the quiet instant and sometimes in a huge crowd. I do not think either instance can be too carefully planned. Rabbi Hanina said, "I have learned much from my teachers, and from my colleagues more than from my teachers, and from my students more than from all of them." As I leave the University for this extended leave, what will I learn in the absence of all three constituencies that might, when I return, interest them? Hanina says that to learn I must have a colleague. I must share the classroom with students. But the sabbatical forbids work in the fields: your field you shall not sow and your vineyard you shall not prune. The aftergrowth of your harvest you shall not reap and the grapes you had set aside for yourself you shall not pick.

And so, upon what is this reflection supposed to reflect? The Oxford English Dictionary tells me that reflection refers earliest (ca. 1386) to "the action on the part of surfaces of throwing back light or heat falling upon them." Like the action of the mirror! Depending on the light available, when I stand before a mirror my image—my image of my image—is visible. It is not me that is reflected in the looking glass, but rather an image of me seen through the distortion produced by the material of the mirror and my consciousness. That is not *me* I see in the mirror, but rather, for whatever motivation, it is the me I *want* to see that is reflected. Some surfaces are more apt to render a clearer image, though not necessarily to reflect a more accurate one. And still, through all of the smudges and fingerprints, we see ourselves anyway. We read those books that we hope will evoke our character, and when we are done, our character, we hope, is altered by the reading. There in the mirror, the reflection is, perhaps, already situated in the past. And here in this writing, reflection helps me see what I can think. This writing thinks. There is no future without a past.

Or reflection might be "the action of a mirror or other polished surface in exhibiting or reproducing the image of an object; the fact or phenomenon of an image being produced in this way (ca. 1430)." The notion of heat has been removed from this definition and priority given to vision. Sometimes, the image I see reflected is devoid of heat; sometimes the image is, indeed, chilling. We could recognize much of our world as reflective, polished surfaces: these books, this writing, my children, my classroom ("All my children?" asks Macduff) reflect me. During this sabbatical shall I look at the reflection in the mirror or at the mirror that reflects? Shall I see the one without the other?

Later, in the seventeenth century, the meaning of reflection generalized and referred to "the mode, operation, or faculty by which the mind has knowledge of itself and its operations, or by which it deals with the ideas received from sensation and perception." That was John Locke's idea of reflection, and it reminds me of the Cartesian philosophy that originally isolated me in my own solipsistic cogitation. I suppose the sabbatical time is intended to afford me the time to hermit the mind to understand itself and its operations (again, shades of Thomas Merton sitting alone in his monastic cell contemplating, or of René Descartes cogitating in his well-padded armchair). But if I've learned anything in therapy, it is that the assertion of certain knowledge, even of myself, is certainly an error. Perhaps engagement in some kind of psychotherapy ought to be a part of every sabbatical—I have, indeed, discovered a return to the couch as part of my time.

The Earl of Clarendon says that reflection refers to "the action of turning (back) or fixing the thoughts on some subject; meditation, deep or serious consideration." Of course, I believe that I have always practiced reflection: by my training and by my study and, well, by my inclination I have engaged in the fixing on, and the serious consideration of, my thoughts upon subjects (on literature and philosophy and now, on whatever cultural events occur in my purview) and objects. Of course, I focus only on those subjects I deem worthy of my focus, and there are oh, so many things I direct so little thought to despite their centrality in our culture—football, for example, and opera. Perhaps the sabbatical year will help develop new themes.

William Pinar (Pinar & Grumet, 1976) has conceptualized curriculum as the process of experiencing the notion of authentic self. The process begins, he says, with a "looking back." Pinar refers to this reflection as the regressive phase of *currere*[1], and argues that it is steeped in autobiography. To know ourselves, he states, we must first know the construction of our pasts. Our stance in the present is constructed by the entirety of our lives in the past. The regressive phase of *currere* is the means to begin to attain such knowledge. For many years now I have occupied this space that Bill has opened, and I have taught and practiced *currere* over the semesters and the years. I suppose the research agenda I proposed to the sabbatical review committee was an attempt to look at my past steeped in a Judaism, which had long been overlaid, and even smothered, by the public educational discourses and practices that were steeped in American Christianity. In his wonderful book, *The Battle for Christmas* (which I was able to read during the early days of Winter Break before official sabbatical time had begun), Stephen Nissenbaum writes, "I once decided that Christmas must mean even more to American Jewish children than to its Christian ones" (1996, ix). I think he meant that the magical season that consumes our culture excludes Jewish children even as it has helped to create American children, even Jewish ones. The fascination of Jewish children with Christmas might derive from their exclusion from it. It was not only the language from which Jewish children were excluded, but the whole culture that has been structured about and by this ultimately Christian event. Nissenbaum's research constructed, perhaps, his absent self. I wanted in my sabbatical to recover a language I almost never knew.

Turning back? But what about Lot's wife?

The sabbatical awards the luxury of time to engage in such reflection— but what about the research agenda? Oh, I know I will have the luxury of time to accomplish both, but what shall I do with the children? And what about the heroism of Esther? And what shall I do with these dry bones?

✡

I had begun searching out the silences: where and how did Jewish discourse get eliminated from the educational practices in the Western World, and more particularly, in the United States. This was a particularly apt subject because I live in *galut*—in exile. The silences deafened me in this strange land. In the part of the United States where presently I live, there are seven Lutheran churches and five Jewish people. In the part of the country from which I derive—New York City—everyone speaks Jewish, even those who are clearly not so; e.g. every employee at every bagel counter in the city knows what a "schmear" of cream cheese means. But in the bagelries of St. Paul/Minneapolis, no attendant seems to have the vaguest idea of what a "schmear" consists, nor very much seems to care. I mean, out here in the Mid-West, they have sun-dried tomato bagels!

I have, of late, become quite aware of my family's separation from the mainstream Christian community and of the mainstream community's separation from us. My daughter, Emma, once returned a video on the morning of the second day of Passover, which happened to be coincidentally Easter morning. When she returned to the car, she announced that the clerk had wished her a Happy Easter. I asked her how she had responded, and Emma reported uncertainly that she had said nothing. She added that she would have wished the clerk a happy Passover, but that probably she wouldn't have known what Emma was talking about. Probably, Emma was correct. The major Jewish festivals, Rosh Hashanah, Yom Kippur, Sukkoth, Passover, Shavuot, are not listed on the school calendar. At the Winter Sing, there are no songs celebrating anything other than Christian winter. Deck the halls with boughs of holly on silent nights. In the Spring, the beginning of the annual candy fund drive is accompanied by a letter of encouragement from the Parent Teacher Organization: "Kathryn Beich, Inc. is the company you can order from. This company offers a variety of foods and collector gifts. Remember, Easter is coming!" Indeed, how could I forget. Rabbi Gordon, whom Emma had approached during Purim as one of her earliest customers, asked if the candy Emma was selling would be delivered during Pesach when *chametz*—leavened food—is forbidden in the house. He was concerned that he not be responsible for our having chametz in the house because of his order. We assured him that the chametz would not be found in our house during Pesach. I think he purchased the Gourmet Nuts.

Our presence in this community *as Jews* is absolutely invisible. A friend once advised me that he had long ago renounced Christianity, that it played no part in his life, and that I threatened our relationship by my public display of Jewishness. I had begun to wear a *kippah* (a skullcap) as part of my regular dress. I responded that even if he renounced Christianity, the average citizen would always assume that he were Christian, but if I renounced my Jewishness, the average citizen would always assume that I were Christian. I have worked in the public schools for almost three decades and not spoken as a Jew because I did not have, and was not offered, a language with which to do so. That is, I was taught as the Judeo-Christian world (their term) taught me to teach. I used the texts I was given, and they were always Christian texts. I did not know that there were alternatives. As Leslie Fiedler reminds me, those "texts taught as 'our heritage' were really their heritage, which is to say, Christian and therefore, necessarily anti-Semitic." (Fielder 1991: 16) I had learned to speak, and even to love, a language that not only excluded me, but gave voice to hatred of me. Now I help others learn how to become educators, and I discover I am using the language with which I was taught. Perhaps it is time to learn a new language. Abraham Joshua Heschel (1955, 24) has said that

> What we mean by the absence of the Bible in the history of philosophy is
> not references or quotations; scriptural passages have occasionally found

admittance. What we mean is the spirit, the way of thinking, the mode of looking at the world, at life; the basic premises of speculation about being, about values, about meaning. Open any history of philosophy. Thales or Parmenides is there; but is Isaiah or Elijah, Job or Ecclesiastes ever represented? The result of such omission is that the basic premises of Western philosophy are derived from the Greek rather than the Hebraic thinking.

Education in North America has been forever based in Greek idealist thinking, and I have spent my life in school. The texts taught as *our* heritage were really *their* heritage. The grains I harvested were not those I had planted. I planted and harvested in fields that did not belong to me. Nor were the tools I used my own.

Perhaps this sabbatical could be a time when I might attend to my work in the fields while letting the fields themselves celebrate the sabbatical rest. I might now consider the connection to my fields when I am not in them—and thereby reestablish a relationship with them—to gain some sense of ownership to the fields and the labor in which I engage within them. When I return in the Fall, the fields will still be the fields, and they will need plowing and pruning and care—grains will yet be grains and grapes will still be grapes. But me, it might be *me* who has changed—and therefore, my relationship to the fields will have changed; thus, the fields will, in fact, have changed—whether they know it or not, and even if they do not care.

I had begun this sabbatical by attempting to add to Western curriculum discourse a particularly Jewish voice (mine particularly) by using particularly Jewish texts—like the Talmud, like Maimonides, like Jewish philosophers who had embraced rather than disavowed their Judaism. I wanted to talk about curriculum in a Jewish voice and register. I had made several presentations at conferences, and I published as often as I was permitted to, using the language I had re-discovered to talk about what I do, about what we do, to talk about my children and other people's children and the schools they attend. I was filled with doubt. As my friend Peter Appelbaum reminds me, "Like much philosophy, it may 'seem' obvious, but you are working through the issues with a new vocabulary and thus a new argument, and therefore, a new context. Also," he added, "things always seem obvious to 'us' once we have worked with them."

Peter has ever been a willing ear to whom I voice not a few of my doubts. It is to Peter I reveal my intellectual insecurities, and his statement was a response to some writing I had done using Talmud as a stimulus to thought, to consider the physical space of the classroom. The Rabbis have a great deal to say about physical space and the activity that may ensue in it. As I continue my research plan during this sabbatical, I am yet thinking about the classroom, but attempting to apply a new vocabulary to it, and thereby make a new argument for it and provide a wholly new context to the discourse about it. How can I not think about school when every morning my children go there? And, in the Fall, I must return there having enjoyed a sabbatical leave from there. Peter is correct, of course; I am working through issues with a new vocabulary, and thus making a new argument, and therefore, constructing a new context. But I thought the land was to be given a Sabbath—no work was to be permitted on it! What will it mean for me not to go into the classroom, not to go to the office, but to reflect upon it? What will I do with the gift of time I have been given?

The sabbatical year is so named because in Leviticus 25:1–4 Adonai commands that when the Hebrew People would come into the land Adonai has promised, in the seventh year that land shall observe a Sabbath rest. During this sabbatical, I must not work my fields. What Abraham Joshua Heschel (1951:10) writes about the weekly Sabbath applies equally, I think, to the Levitical sabbatical:

> The meaning of Sabbath is to celebrate time rather than space. Six days a week we live under the tyranny of things of space; on the Sabbath we try to become attuned to holiness in time. It is a day on which we are called upon to share in what is eternal in time, to turn from the results of creation to the mystery of creation; from the world of creation to the creation of the world.

Traditional Sabbath observance forbids the thirty-nine kinds of work engaged in when the sanctuary was being built. If God ceased the work of creation on the seventh day, then humankind should stop creation on the seventh day as well. In contemporary terms, Sabbath observance forbids the transformation of anything from one thing to another. On the seventh day, Heschel (1951:31) writes, "man has no right to tamper with God's world, to change the state of physical things." Thus, Heschel's notion of the Sabbath is a celebration of time. Zahorski (1994:5) notes that Pliny the Elder recounts a Jewish legend featuring an "ancient river in Media named Sabbation, which flowed for six days but rested the seventh." Nature had ceased its natural courses on the Sabbath. It is to this story that Zahorski attributes the idea of the sabbatical. But it is not my fields that will cease during this sabbatical; rather, it is I who must cease working in the fields, and then they too will rest. And if the fields must be left to rest during this sabbatical leave, must I then abandon the agenda I had so carefully planned almost sixteen months ago? Or does the sabbatical mean that I must abandon its agenda and retain the activity? Upon what should I reflect? Upon what objects should I fix my thoughts? Then how should I begin? And how do I proceed?

According to Leviticus 25: 1–7, the sabbatical is a rest for the land. This is not uninteresting because it is clear that the earth itself does not cease to move nor produce growth during even the sabbatical year. But Leviticus specifically abjures all but subsistence use of the aftergrowth of the vineyards. Much of humanity's metaphors derive from the cycle of the seasons: a year never goes by that somewhere the earth doesn't move. So this sabbatical year of the earth is a year when humans are to let the earth for a year move to its own rhythms—to not till it. To give the earth a Shabbat rest. Now the laws concerning Shabbat are immense—and immensely important. The transgression of some of those laws calls down even the death penalty upon the violator. Our treatment of the earth and all that rests therein has called down the death penalty for so many of us for so many years now: Love Canal, Chernobyl, Hiroshima, Vietnam, Bahgdad, New Orleans, Faluja. This is a clichéd list—every day I discover new misuses and abuses of the earth.

So if the land is to rest—to be allowed to rest—then what does everyone who worked the land *do* during the Schmettah? One must assume that Leviticus refers to an agrarian economy when most people worked the land either for themselves or for their employers. Though the fields be replete with food ready for harvest, no harvesting was permitted in the fields. That is, the sabbatical rest for the land forbids any activity in the fields that would change them from the state in which they were left at the end of the previous year. It interests me that Leviticus is phrased thus: "the land shall observe a rest"—the living earth. David Abram's book, *The Spell of the Sensuous* is about just that idea. The book argues that we have severed our relationship with this living earth—that we have become separate from it and lost a nourishing and sustaining relationship with the earth. The Hebrew alphabet, which contains only consonants, reminds us of our once central, and now lost, attachment to the earth. Without the breathing that is the vowel sound, the letters may not speak and, of necessity, must remain inert. Without air, all life perishes.

> The invisible air, the same mystery that animates the visible terrain, was also needed to animate the visible letters, to make them come alive and to speak . . . The letters themselves thus remained overtly dependent upon the elemental corporeal life-world—they were activated by the very breath of that world, and could not be cut off from that world without losing all of their power. In this manner the absence of written vowels

ensured that Hebrew language and tradition remained open to the power of that which exceeds the strictly human community—it ensured that the Hebraic sensibility would remain rooted, however tenuously, in the animate earth" (1997: 242).

I think this is what Arthur Waskow refers to when he utters the sacred name of Adonai—the tetragrammaton—as the sound of a breath: YHWH. In a midrash to the story of the burning bush, Waskow portrays Moshe asking God for a name by which God might be identified. God admonishes Moses not to try and pronounce the name of God; its pronunciation is not articulation but respiration. I have heard Waskow in prayer breathe the name of God as the statement of God's name. It is the substance of life itself. God says to Moshe, "I am the Breath of Life, and the Breath of Life is what will set you free. Teach them that if they learn my Name is just a Breathing, they will be able to reach across all tongues and boundaries, to pass over them all for birth, and life, and freedom" (Waskow & Berman 1996). During the sabbatical, can I learn to breathe again?

Leviticus says that the land is to observe a rest. Commanded not to do anything to change the land, humans are to experience an opportunity in time to reestablish a relationship with the land based not on its exploitation but on its celebration. The land's sabbatical precludes any exploitation that derives from its uses, precludes the use of all that we have come to associate with progress and the advance of humanity. Heschel's notion of Shabbat is invoked here: "The Sabbath," he writes (1951:31), "is more than an armistice, more than an interlude; it is a profound conscious harmony of man and the world, a sympathy for all things and a participation in the spirit that unites what is below and what is above." On Shabbat, Heschel writes (1951, 32), we might "rest even from the thought of labor." But how shall I celebrate the sabbatical leave without thinking of my labor—of the school. Everyday my children go there, out into the fields, as it were. In the Fall, I must return there myself. My entire research plan depends on my thoughts concerning my labor. Heschel (1951, 31) tells the following story:

> A pious man once took a stroll in his vineyard on the Sabbath. He saw a breach in the fence, and then determined to mend it when the Sabbath would be over. At the expiration of the Sabbath he decided: since the thought of repairing the fence occurred to me on the Sabbath I shall never repair it.

But perhaps I am not so pious a man.

✡

Torah says that one can eat of the aftergrowth of the fields but make no other use of them: whatever grows—rather, whatever has thus far grown, may be eaten but may not be used for any other purpose. What could that mean? Obviously no wine may be made from the grapes—you cannot profit (reap?) from the aftergrowth. You may not sell the produce from the aftergrowth; that is, you may not harvest the surplus growth—presumably that which grew after the last harvest and as a result of unseasonable weather (there is yet the earth that moves regardless), or what results from special plantings (like late peas and beans in our gardens); you may not prepare the fields for the next year's growth and harvest. Furthermore, this aftergrowth and remaining produce may not be selfishly possessed but must be finally made available for everyone in the household—the slave and maidservant, the animals and the worker and "your resident who dwell with you." From my present perspective down here in the basement, I wonder who might that "resident" be? There is a great deal of surplus growth in certain quarters in the United States. Perhaps this sabbatical command ought to serve as an example during these days of the evisceration of social programs. Down here I have heard that one-third of all American children live in poverty despite the activities of two working parents.

Perhaps to be true to the Biblical sense of the sabbatical, what I might do is keep a record of what occurs during this leave—("Do you keep a journal?" he asked.[2])—it can be a matter of reflecting upon the papers I have already written, of my family life, and of the books that I have in the past read. This journal might consist of the materials that comprise my day during the celebration of the sabbatical. And then, at its end, I might consider a means to organize these fragments into a person. I have never worked without stacks of reference materials gathered from the fields. The sabbatical commands insist that the fields be left to rest!

Would I have been awarded sabbatical leave if I had written that its agenda was to be immediately abandoned? Zahorski (1994, 31) writes, "Most sabbatical programs demand thorough and detailed proposals which are then judged on comparative merit." How might the sabbatical leave committee have responded if I had told them that any pre-set agenda would defeat the whole rationale for sabbatical leave? Would I have been awarded a sabbatical leave if I told the committee that I meant to be reflective and think about my life? Focused research in the luxury of non-obligated time is wonderful, and for many this is the business of the sabbatical. But I am reminded that those fields experiencing the sabbatical are not to be touched—oh, yes, they may be gleaned, as it were (though not by me!), but not plowed, nor prepared, nor sown, nor pruned, nor sold. The aftergrowth of the harvest I may not reap. Does that mean that no articles may be submitted? That would be a relief. Interestingly enough, the sabbatical of the fields gives rest to all—I wonder if my colleagues are relieved that I am not on campus? Or that some of the editors with whom I deal will not hear from me this year? When I make my report, what shall I answer for the leave?

There will be separation at the end of this sabbatical—reentry into the university will be difficult, I think. Jewish practice provides for this exigency—Havdalah marks the separation of Shabbat[3] from the other six days, demarcating the holy from the secular. During Havdalah we extinguish the braided candle in the cup of wine and place some of the liquid on our eyelids. Then, it is hoped, we will throughout the week see the world through the glaze of Shabbat. During Havdalah we sniff aromatic spices to hold in our senses the sweet smell of Shabbat. But what ceremony separates the sabbatical from the previous six years? What ceremony will separate the sabbatical from the next six years?

And so I think continuously about this sabbatical. Where and what are my fields? I am a professor in a mid-Western University and I work in the fields of teacher education. I teach curriculum classes. I teach Introduction to Curriculum, Methods and Assessment and Foundations of Education to future teachers of America. My fields in this sense may easily lie untouched. I don't even go into the office to retrieve my mail, and certainly I don't go into the classroom. Of course, the classrooms go on even without me, but they are changed in my absence.

I am being intentionally arbitrary (that is not the word, I seek—the English word for which I search means "nitpicky," means 'pilpulish,' the latter a Hebrew word referring to a type of argumentation—dialectical reasoning). And so I return to the original question. What am I doing? What is permitted to be done? If all I do is take what I have done in the office and bring it home, then I am not fulfilling the commands of the sabbatical year. If I am not permitted to do what I did during the year— if, indeed, I am—to use the parlance of the last few years, supposed to "give it a rest," then what am I supposed to do with myself? (My cousin by marriage, Ricky, tells me that where he lives a sway of a limp wrist with the declamation "but we're not going to go there!" is an indication of a field not to be ploughed, is an indication of a field left to celebrate a sabbatical, so to speak. All of this cogitation about what to do on sabbatical consumes the sabbatical—perhaps I shouldn't go there).

This is not an uninteresting predicament given that people keep inquiring into my activities. Given that *I* keep inquiring into my activities. How does a person such as myself, how does a person who is myself, celebrate a sabbatical year? Should I read only materials that have nothing to do with my subject matter per se. That is impossible. There is, I am afraid, nothing that isn't finally part of my subject—curriculum. Should I stop reading? That is impossible: reading is akin to breathing. Perhaps I should only re-read books I have completed in the past? Should I retire into the monastic environment? That is impossible; I have two wonderful children with whom I would regularly live. Should we have taken off to some exotic far-away place? That hardly seems an appropriate activity; certainly we could not have paid for it either materially nor emotionally—how could I have stood being away from here for any length of time more than a weekend? Away from my books. Daniel Isaacson, in E. L. Doctorow's *The Book of Daniel* (1971, 13) sits amidst the stacks in the library and is consoled. "I feel encouraged to go on," he announces. How could *they* have lived without their friends and their fields? My sabbatical does not permit me to uproot everything—indeed, I am forbidden to change the fields.

NOTES

1. *Currere* is the verb form and means "to run." *Curriculum* is the noun and refers usually to the course around which one runs.
2. This is the first line in the first volume of the multi-volume journals of Henry David Thoreau. It is the question posed to the recent Harvard graduate by his mentor and friend, Ralph Waldo Emerson.
3. I use throughout this book the Hebrew form of the word for the seventh day.

CHAPTER 3

THE THIRD SAPLING

Leviticus, the third book of the Torah, begins with the rules and regulations concerning the ritual practice of the sacrifice. Sacrifice was the primary process by which the human population established communication with the divine. Through the conciliatory act of the sacrifice, human beings could initiate with divinity the illusion of a sustaining and cooperative relationship with the earth and all its inhabitants thereon. By communicating to God, placating and assuaging and offering thanks to God in the practice of the sacrifice, human presumption could assert some measure of control over the exercise of their daily lives. The Levitical code established the means and measures by which designated sacrifices must be ritually and legally carried out. Specifically, Leviticus identifies the guilt offering, the meal offering, and the peace offering as categories of sacrifice for effective conciliation with God. In explicit detail, these early chapters mandate not so much the whys and wherefores—for exactly which specific act a particular sacrifice must be made—but the hows and whens and whoms—by what procedures and with which participants the sacrifice must be accomplished. Indeed, it is interesting to consider that even before the moral code is explicitly stated in the later chapters of Leviticus, the Torah establishes the detailed *structure* of the sacrifice that mitigates that behavior. The means of repair are delineated before the rent is made. Hope does not replace despair but makes it unnecessary.

And the Levitical author speaks not about specific human acts, but in terms of general behavioral patterns. That is, the text prescribes the explicit procedures for the sacrifice in the case when "an individual person from among the people of the land shall sin unintentionally . . ." I wonder who could ever be exempt from such a commission? I think that by providing means of expiation without defining the specific transgression, the authors of Leviticus demand that there be culpability for our actions but not hopelessness for our transgressions, and insist that there be responsibility for our deeds and yet the presence of mercy. Leviticus acknowledges the inevitability of humans to sin and yet offers the assurance of redemption. As a result of the human condition, human beings feel, perhaps, a need for penance and hope; the sacrifice offers the people this sustenance. The sacrifice is constructed as mediation between the divine and the human; the sacrifice serves to facilitate atonement for wrongdoing and to render devotion for beneficence. The sacrifice ensures the presence of the future.

The existence of the sacrifice derives from the existence of a standard for human behavior to which we may aspire. Because structures are established to mitigate as yet unnamed human failings, as well as to offer thanksgivings for apparent divine blessing, we may, perhaps, assume that the exemplary life epitomizes holiness—a life of unquestionable righteousness—as the goal for human behavior. If it is finally in expiation for acts committed that the sacrifice is mandated, then it must be in acts that righteousness might be sought. Heschel says, "It is in *deeds* that man becomes aware of what his life really is, of his power to harm and to hurt, to wreck and to ruin; of his ability to derive joy and to bestow it upon others; to relieve and to increase his own and other people's tensions . . . In his deeds, man exposes his immanent as well as his suppressed desires" (1959, 82). Leviticus offers the people a way to return to paths of probity and virtue from patterns of behavior that are antagonistic to that goal. Leviticus establishes categories of the sacrifice based on an ethical code that has as its goal the achievement of holiness. It is, after all, in Leviticus (19:2) that God commands the people of Israel: "You shall be holy, for holy am I, Adonai, your God."

And because of their generality, the broad sacrificial categories enunciated in the opening chapters of Leviticus inspire and require personal reflective contemplation and assessment regarding the exercises and motives of the *practices* of daily life so as to ensure the necessity for, and the correct performance of, the appropriate sacrifice. We are made responsible in our freedom. And the sovereignty of the monotheistic God of the people of Israel ensures that a uniform and consistent standard of human activity be maintained. There are no personal gods with whims and individualities and peccadilloes who must be individually becalmed;

rather, there is in Leviticus 11:44–45 a single, non-corporeal Divine Being to whom all are responsible and who is responsible to all. To be holy is the obligation God ascribes to human beings. "For I am Adonai your God—you are to sanctify yourselves and you shall become holy, for I am holy . . . For I am Adonai who elevates you from the land of Egypt to be a God unto you; you shall be holy for I am holy." I think that the detailed description of the prescribed sacrifices even before the enunciation of the explicit moral code assumes the enduring failure of the people to attain holiness and the yet abiding quest to aspire toward it. The first five chapters of Leviticus describe the sacrifices that must be uniformly brought by all of the people in obeisance to a single God. The first five chapters of Leviticus establish a standard of behavior not by naming the specificities of that behavior, but by defining the categories of sacrifice that may mediate it.

Now, once holiness is established as a goal and procedures to mitigate and cleanse sin are established, then it is necessary for the designated priestly intermediaries to be instructed as to their precise role in the sacrificial rites. And because lapses in human behavior were already expected, then the purity of the priests—the mediating classes—must be ensured and their behavior authorized so that the prescribed sacrifice that they will carry out will be acceptable. The priests must be given directions. And so it interests me to note that the language regarding the sacrifices, which opens Chapter Six—the directions *to the priests*—represents an interesting linguistic variation from the opening language of the preceding five chapters—the direction *to the people*. There, the Torah had stated simply, "And the Lord spoke unto Moses, saying, '*Speak* to the Children of Israel.'" But Chapter Six begins, "And the Lord spoke unto Moses, saying '*Command* Aaron and his sons.'" The question is asked, why do the priests need to be commanded? After all, if it is the priest's place in society to ensure the potential for holiness in the performance of the sacrifice, then why must the priests be here commanded to do so? The scholar Rashi (1040–1105 C.E.) says that the command to the priests was necessary because the sacrifice referred to was a total one; that is, there would be at its end nothing left over for the priests. Therefore, their participation had to be commanded because they would be less apt to take part in an activity from which they would receive nothing for themselves. Rashi reads back into history human self-centeredness. That is one answer.

But I learn from Rabbi Morris Allen another reading, which I prefer. Here it is: The direction to the priests comes with no explication justifying the carefully detailed structure of the sacrificial rite. For each specific sacrifice, a procedural order is mandated and the text simply directs the priests that it is *this* they are to do. These directed acts are not to be questioned; the priests should expect no rational explanation for the particular formulas. These measures so carefully set out are simply to be performed. As holiness is the goal—"As I am holy so you shall be Holy"—and as it is the sacrifice that renews the possibility for holiness and as it is the priestly caste that must perform this rite, then there must be no opportunity for innovation or option in the sacrificial rite. These things *must* be done. Command the priests: these things you must do.

So things might have remained. But the destruction of the second temple in 70 C.E. destroyed the central framework of Judaism. The Torah had situated all performances of rites and ceremonies in the Temple—including the sacrifices—and to be performed by the class of Priests whose position was inherited throughout the generations. The loss of the Temple eliminated the possibility of the sacrifice by removing its sacred center; when the holy space of the temple altar was destroyed, the possibility for sacrifice also disappeared. The priestly caste too was now rendered unnecessary in its historical and ritual form. Without the Temple, there was no place for the sacrifice; without the sacrifices there was no need for a priestly caste; without the sacrifice, the people had nothing to hold them to the standards of holiness the Torah had chosen for them. If Judaism was to survive, then it had to be reinvented.

Such was the accomplishment of the rabbis in the years following the destruction of the Temple. The Rabbis reconfigured Jewish practice and reinvented Judaism by reading into the Torah precedent for situating the practices of Judaism *within the home* rather than within the Temple, and they gave responsibility for its practices to the people rather than to the priests. The rabbis replaced sacrifice with prayer and study, the former to ensure holiness and the latter to ensure continued obedience to the eminently interpretable text. That text, the *Tanach* (an acronym for the twenty-four books of the canonized Bible—the Torah [the five books of Moses], the Neviim [the books of the prophets], and the Ketubim [the writings set down and not spoken as prophecy]) portrays the history of the Jewish people as the revelation of God's will and the continual falling away of the people from accomplishing that will. The *Tanach* tells the story of the establishment of holiness as a social and personal ideal for the Jewish people and of the constant pursuit and eternal falling short of that achievement. To preserve Judaism, the rabbis created a Torah that required interpretation based upon this teleology of holiness and the Rabbis offered a variety of hermeneutical principles to accomplish this exegesis. In its form and content, the Talmud, the foundational document of Rabbinic Judaism, reveals to us that interpretation is never exact or final. Our understandings are never complete; the quest for knowledge is unending. The quest for holiness is forever unfinished because holiness—like learning—is a pursuit and not an achievement.

However, with the enormous growth of our knowledge of the universe and the development of our sciences and technologies, the human being has more and more come to believe in her/his power to hold dominion over all of Nature. We stand in the world as its master. The quest for holiness has been replaced by the development of systems of authority and power. That ascendancy of human control over the universe has been attributed to the powers of reason—the workings of the mind. Thus, the greater our control appears, the more power we ascribe to our reason. We have in our beliefs become the absolute creators of the physical world. Indeed, as we learn to create and direct life, genetic science has made us appear as godlike as the God of Genesis. Scientists have come to believe that their knowledge has become authority. Former President Clinton, standing in the East Room of the White House before a first survey of the entire human genome, remarked: "Today, we are learning the Language in which God created life . . . With this profound new knowledge, humankind is on the verge of gaining immense, new power to heal." Doctors accept now, too easily perhaps, that they control—and will continue to control, wellness and disease. We have come to believe in a direct connection between our behavior and the effects of it. We have come to believe in a direct connection between our behavior and the effects of it.

And teachers, of whom I am one, assume the power to acquire requisite knowledge and to acquire, as well, the capacity to distribute that knowledge to people and to make them learn. We too readily believe that the development of the correct standards and the insistence on their acquisition will ensure the nation's greatness and continued ascendancy over the world. There are committees writing standards in almost every state of the Union; there are committees in Washington charged with the establishment of national standards for curriculum. If only we could achieve these academic standards, then how magnificent would be our works!

And so perhaps it is the language of Leviticus—the command to the priests that it is this they *must* do— reminds us that the priests, now become scientists and doctors and rabbis and teachers—are perhaps not the final cause of anything and are not the final authorities of the effects of their causes. There are greater goals than those we set or even that we could know; perhaps it is this humility that should underlie our acts as educators. "You shall be holy for I, God, am Holy" (Leviticus 20:26). Leviticus humbles the priests by commanding that some things must be accomplished without question. "Speak to the priests, *command* them . . ." It must have been, I think, a chastening experience. Finally, the priests are made to accept that they are instruments of a greater goal and not the primary causes of its achievement. In Torah, the greater goal—holiness—is finally not a human construction but a human aspiration. There are, perhaps, standards that we cannot set, but to which our practices must aspire. The achievement of holiness requires that there is more to the world than can ever be understood.

In Leviticus, God demands a standard of ethics—*I am holy and you shall be holy*—but by the establishment of the sacrifice concedes that humans, left to themselves, would never achieve that state. The priestly caste is thereby created and empowered to ensure that the paths to holiness are forever available. But the command, which is made to the priest at the beginning of Chapter Six, announces that finally, ultimate authority is always outside the human capacity for knowledge and control. Holiness is not humanly circumscribed but divinely ordained. Since its definition is beyond the priest's ken, then certain acts must be performed without question simply because they are commanded.

But we must not think that these commandments are humanly devised or subject to human exigency; nor must we accept that they are beyond human capacity to accomplish. I think that this command—this you must do—requires no more than that we stand in our classrooms face to face with each other: "You shall love your fellow as yourself" (Leviticus 19:18). If we have replaced the redeeming sacrifice with study, then it is this we must do. Educators must remember that though we lack the certainty of the ends of our actions, it is still by our deeds rather than by our intentions that we might be measured. It is by our stance in the classroom, rather than any prescribed measurable objectives, that we are ourselves to be measured. I become a teacher when I engage in the attainment of holiness and not when I organize the classroom about objective standards. I become a teacher when I acknowledge my responsibility before those whom I face. Not less than thirty-six times are the Israelites charged to care especially for the stranger, for once they were strangers in Egypt. Who else are all those people in the seats before us but strangers? We must never assume to treat them as less than strangers. The image of God is best honored in the right given the stranger—the students—than in any symbol. We might remember that when next we issue standards and grades. I think it is important to recall that when we think we are certain, we are certainly in error.

The present in which I exist still sets holiness as its goal, but I think that that holiness is very much identified with learning and with education. It is the promise of education that the world be saved. More particularly, it is the promise of education that the ascendancy of the United States be continued. Even more directly, it is the promise of education that the private fortunes and happiness of all be increased. It is the promise of education that we create the Great Society. The generalized categories of the Biblical sacrifice established ethical standards to ensure that the quest for holiness be unending and that the priestly mediators remain humble. Perhaps it was the order to the priests that it was this they *must* do that remembered for them that it was human and not priestly activity that achieved holiness. The purpose of the sacrifice was to reassert the goal of holiness rather than to confer it. Holiness yet required human deed. The deeds for which sacrifice were to be made remain in the Bible unstated so that the pursuit of holiness remained a process and not a product; Judaism established the achievement of holiness in the state of action and not in the state of the soul.

The issue of holiness is, indeed, one of standards, but they are standards of ethical behavior and not of material acquisition or power. Holiness is an ongoing endeavor and not a *fait accompli*. Heschel writes that man is in need of bread, but bread is not in need of man. Justice, however, needs to be pursued and therefore, justice is in need of man. Justice requires human action. So too, with holiness—holiness must be actively sought and not ideally conceived. Holiness, of which justice is a component, demands human pursuit. Holiness is a mutual relationship based on obligation. Heschel writes (1955, 291–92), "The sense of obligation expresses a situation in which an idea, as it were, is waiting to be attained." The language of Leviticus suggests to me that because education, too, is a quest and not an achievement, then the emphasis on its products attenuates its purpose and the quantification of its goals falsifies its design.

Indeed, as education is an obligation—it must be pursued and therefore, requires human activity—education is a commitment to the other, a demand upon and not a conferral. Education must be measured not by what the educator prepares in the effort to achieve formal academic standards, but by the degree to which response is made to the student's demand. I can only be *I* when no one can take my place before the unspoken demands of the other. "I am *I* as if I had been chosen" (Levinas, 1994, 35). I become the teacher when I accept this obligation. The language of Leviticus suggests to me that what the teacher needs is not standards but students. And as we are all someone's teachers, so are we are all students to someone. The Levitical command humbles the priest even as it is the student who humbles the educator: this you must do! How might we be reminded that too much of what we do is beyond even our understanding! How can the educator come to know that our powers are very limited! How defiled our standards seem when their existence requires the silence of those who call us to our service; how puny our achievement seems that it can even be measured! If, as educators, we would be reminded that our knowledge is limited, then our hubris in the classroom must vanish. How would our classrooms change when the certainties of standards are called to question and the demands of our students are given ear?

Education must answer to this obligation—to holiness. I think holiness abjures injustice. Standards are human creations tainted by injustice.

✡

Social justice and equality—well, these are standards that extend beyond the classroom, but that require the classroom for sustenance. I do not know how they might ever be quantified. Once you know the properties of an isosceles triangle, you can figure out the height of a flagpole by measuring its shadow and angles. Do we spend as much time in school discovering the heights of a person's character? If a train leaves New York at 60 miles an hour heading west and another train leaves Los Angeles heading East at 90 miles an hour, at what location will their paths cross? Do we wonder as much about potential meetings between classes, or between ethnicities, or between family members, or between teachers and students heading rightfully in various directions? Presently, every achievement in school is scored and measured and evaluated. Every act is reduced to a quantity. Our obligations are actually blunted by education. Our hands remain always clean of the world. In Leviticus we learn that when you reap the harvest of your land, "you shall not complete your reaping to the corner of your field, and the gleanings of our harvest you shall not take." These leavings from the corners of the fields are available to the poor, to the widow and orphan. These gleanings represent our obligation to the less fortunate. But how big or how small might that corner be? We learn from the Rabbis in Pe'ah 1:1 that, in fact, the corner may not be measured: "These are the deeds for which there is no prescribed measure: leaving crops at the corner of a field for the poor, offering first fruits as a gift to the Temple, bringing special offerings to the Temple on the three Festivals, doing deeds of loving-kindness, and studying Torah." There are some things, the Rabbis suggests, that may not be quantified; two such matters include our obligations to the other and our responsibility for

continuous study. Their achievement is unending; our pursuit of th.....
measures our holiness. As I have said, one of these unmeasurable practices, the rabbis state, is study. Ironically, our national government and
many of our states presently mandate such measure.

Because daily life is so complex and filled with such exigency, the rabbis suggest that the achievement of holiness is a process that must be continually renewed; hence, the necessity of the sacrifice. Holiness is a stance
in the world and not a state of the individual. I think the command to
the priest reminds the priest that it is not his performance that effects
holiness; the priest is merely an instrument in the process. The priest
serves purposes that s/he can never know.

So, too, does the educator. With the loss of the sacrifice, I have
become the priest. I have been assigned as mediator between the achievement of holiness—this time defined by a measure of learning—and the
present. As education enables the individual to dwell in the world with a
sense of power and potential, as education assures us that our failings
today will become attainments tomorrow, as education offers renewal of
body and spirit in the interrogation of present questions and the cleansing promise of redeeming answers, then I am become priest who handles
the present to ensure a future. The conditions and questions that are
brought to the classroom are the result of the lives lived outside of them.
If I am to serve the demands of the students, then how can I presume to
speak first and with authority? Who can speak authoritatively of their
students' lives that a curriculum could be constructed and standardized
to adequately respond to their unspoken demands? I teach the purity of
mathematics to the child who will slaughter his classmates with the
assault rifle no one knows he possesses. I offer the clarity of science to the
child befuddled by a bitter divorce. I insist on my narrative of history to
the child who is silenced from it. I demand the correctness of a grammar
to a child whose parents would not understand it and for whom it serves
little purpose. If the educated person takes a stance in the world, which
heals that world—as the quest for holiness sought repair for the injustices
of the present—then what ought a proper educational goal be? Education
should be understood as the means to intellectual and spiritual elevation.
For these things, there must be no measure.

Education should ensure that not only the material but also the
inward life of the individual be developed. Education should address not
the isolated intellect, as the advocates of standards suggest it ought to do,
but the hopes and dreams of the self of which intellect, the complex,
reflective self is merely a part. That self should have the opportunity to
experience the world in ever widening perspectives, to be free to reflect

upon that experience in freedom. Education offers redemption, but not in an absolutely material fashion. A perfect report card must speak to more than a five-dollar reward or admission to a prestigious college. A perfect report card may not be written but must be lived. Education must offer the opportunity for constant renewal. It must acknowledge that the only constancy is change. It cannot hold to standards today that will be obsolete tomorrow. Only holiness may be standardized—the quest for a perfection that is immeasurable—*this you must do*—though its definition remains beyond human ken. "Speak to the priests: Command them!" Education is a private engagement in a public world for the redemption of both. Education should obligate us to pursue justice—justice requires human action. Our education should remind us that something is demanded of us.

Too often, however, I think we teach how to listen and direct what to say. Sometimes I think that in the schools we have ceased standing in awe of the world that surrounds us. I think rather, that the world terrifies us at times and our standards and certainties assuage our fear. Our presence should offer the possibility of redemption and not its purchase.

Sacrifice is no longer a social practice—though I despair that we too readily are prepared to bind our children. In the school, our actions should be governed by the social vision of its inhabitants. Education should be the establishment of the means to ensure the achievement of our social vision. Leviticus suggests that, for the priests, there is an order of reality that is not questionable but must be simply obeyed. That order entails holiness and since holiness in the Torah is of the nature of God, its achievement is ultimately inscrutable. No wonder the priests may not invent their own methods! In our modern world, education has replaced the sacrifice as the means of achieving a social vision. But what social vision could lie behind the achievement of stringent, and often punitive, standards? Education should be built upon the vision of a just world. We must let our students learn that there is no measure for the corners of our fields and there is no end to study. But at present what holds us educators to our purpose now is the achievement of quantified standards.

One secular version of Leviticus was stated perhaps by Horace Mann: "Education is to inspire the love of truth, as the most supreme good and to clarify the vision of the intellect to discern it" (in Messerli, 1972, 50). I can accept the Protestant Mann's clarion call and reject his insistence that the way to truth may be definitively known. I can accept the goal and yet acknowledge that there is more to truth than the intellect can know. The teacher mediates between the goal and the reality, between the ideal of truth (I wonder if Mann distinguishes between the love of truth and the embracing of it) and life's daily exigencies. What the teacher ought to fashion by education is a moral and spiritually sensitive individual who possesses a strong sense of duty to others. If the acts of the Hebraic priest in the sacrifice ensure the possibility of the continued quest for holiness, then in our secular society the teacher assumes responsibility for maintaining allegiance to the love of an indefinable truth despite the present lack of vision of the occupants of the class. If, via the sacrifice, the Levitical priest mediates between the aspirations for holiness and the reality of sin, then between what exigencies does the contemporary educator mediate? Could it really be the achievement of a set of established and written standards that are quantified and finalized and then discharged or discarded? What ought to be the contemporary goal for education that the teacher's work might facilitate?

For Leviticus, holiness is forever ineluctable—hence the directives for the sacrifice even before the statement of the moral code, which the sacrifice mediates. And when the Rabbis replaced the sacrifice with study and prayer, they did not imagine that they had definitively defined standards. Truth may be in Torah, but Torah remains always subject to human interpretation. In an extended discussion on the relationship between written and oral Torah, David Kraemer (1990) claims that the Rabbis in the Talmud argue for the equality of the two. That is, the Oral Torah, teachings and traditions ascribed to Torah and given to Moses on Sinai but not written down in the canonical five books, is given canonical status by its equation with the written text. Interpretation, thus, is a human affair that aspires to understand that which cannot be ultimately understood. But we do our best. Kraemer (1996, 32) writes, "Are the meanings there, then, in the text as it is written? If we may discover these meanings and defend them with reference to the text as composed, then they are there." The truth of Torah is a goal but not an end. The Rabbinic tradition prescribed hermeneutical rules by which Torah may be interpreted, but they never seemed to doubt that interpretation was required. In the absence of the sacrifice, prayer and study became a means to holiness. In study, we find the practice of study; in the practice of study we discover the life we want to live. The search for truth is a goal and not an end; in Talmud, the absolute does not exist. But in the writing of national standards, truth is created. We are become God. Leviticus reminds me that this is a dangerous and proscribed hubris.

✡

Leviticus establishes human conduct on the basis of responsibility. I discover in Leviticus not a source for perpetual guilt, but a realization of standards by which holiness might be defined as an ethical foundation for the establishment of a holy people. This holiness espoused in Torah is premised on the reality of a history of oppression and is based on the quest of an ethics, which remembers that history. Sacrifice acknowledges the inevitable contradiction that exists between human frailty and ideal goals; the sacrifice ensures redemption despite present corruptions. Despite what I am today, the structure of study promises redemption tomorrow. Leviticus suggests to me that I must act regardless of tangible reward and regardless of the absence of quantifiable results: the world depends on my actions.

It is an interesting connection between sacrifice and study. If sacrifice is effected as expiation for sins or as acknowledgment of our gratitude, if sacrifice is a means of connection between heaven and earth and if study has replaced the sacrifice, then what am I doing when I engage in study? What might I be seeking to control in the exercise of scholarly work? Doesn't Leviticus announce the impossibility of ever finally knowing fully what I seek? I am always insecure in what I do not know and even in what I do not know that I do not know. Doesn't Leviticus suggest that there are some things over which human beings have no control and must simply do without question in the quest for our contemporary mode of holiness? Since I have chosen an academic life and the profession of teacher, these are not small questions in my lexicon of questions: how I respond to these matters helps define a great deal of my life. How I respond to these matters explain the relationships I have established in this life in this world and at this time.

When I study, what is it that I mean to do and what is to be done?

When I teach, do I make them learn? What are the standards of holiness to which I am obliged as a scholar/teacher? Command the priests: this you must do.

✡

A MEDITATION ON TEACHING

Anna Rose and I sit quietly Sunday night watching Disney's newest version of *Cinderella*. Appropriately enough, they chose to televise this production on Valentine's Day. I mean, this is the day dedicated to love, and we know from the start that the lovely, enamored pair will live happily ever after in heterosexual wedded bliss. (I know that I will finally have to find some way to discuss these exigencies with my now entranced five year old). Normally on Sunday evenings, Anna Rose would be out ice-skating with

her sister and her mother and I, well, I might practice more guitar funda-
mentals and sing a few songs to the audience of four cats. But tonight,
Anna Rose was not feeling that well; tonight she wants to go to bed rather
early. And so she and I stayed home and she sat on my lap in the comfy
chair and we stared at the television screen. When I was a child, every
Sunday evening was taken up with the Walt Disney Show. I grew up with
Spin and Marty and Davy Crockett and Marion, the Swamp Fox. Even
now, knowing what I do about ideology and historical imperialism, those
television moments and stories are yet dear to me. I recall those nights
fondly as I sit here this evening holding Anna Rose securely. I wonder for
whose safety am I concerned? I wonder who really is being held here?

Oh, this present Disney production is not much better than adequate—
multiracial, a few songs I recognize, a few I don't. To my mind the whole
production is eminently forgettable. Perhaps that is how popular culture
works most effectively—insidiously and insinuatively, quietly and without notice. Like those gases the Joker employed in the Batman adventure
series, or that are dispelled by all of the criminals occupying all of the
comic books and comic book programs I have ever consumed. They set
off an innocuous and odorless gas and the innocent bystanders breathe it
in unknowingly and are suddenly overcome and rendered unconscious.
While they are rendered passive, the evil forces change their world. Anna
Rose is quiet; well actually, she is almost asleep, but if I move I might dis-
pel the spell. I remain and watch Cinderella battle with her siblings. I am
interested in what Anna Rose is not seeing.

Tonight's multiracial production of Cinderella is supposed to be a new
twist on an old tale. Anna Rose has seen this story in various forms
numerous times, and this production is just not that interesting. Of
course, this new twist depends on the essentially racist character of our
society—why else notice the multiracial casting?—and we are a household already quite aware of issues of race and ethnicities. This is, as well,
a story that is essentially misogynist—the only appealing woman in the
whole production is Cinderella, whose reward is a wishy-washy prince.
Disney has not created a feminist fairy tale here. Nor could these productions televised on the mainstream Disney network program be seriously
considered *strong readings or potent re-readings* because drastic editing
occurs to protect children from life. For example, Disney's offering of *Mr.
Holland's Opus* edits out all evidence of human frailty and sexuality.
Sometimes, films must be edited to fit the opus into the *de rigueur* two
hours with a great deal of commercial interruption included. These
movies have not been given a sabbatical, as it were. They have not been
left alone; they have been worked in, as it were.

Then suddenly and almost garishly, appears this advertisement calling for nominations for Disney's "Teacher of the Year." It is a cheerful bright infomercial: a schoolroom resplendent in primary reds and yellow and blues—sparkling children bouncing about with ebullient smiles and carefully mussed hair. There, on the screen, is the classroom of my dreams! These should be my students. Hey, this is my contest. I have always wanted to be teacher of the year. Of course, I always imagined that I was teacher of the year, only that the news had not yet traveled very far. In reality, I have absolutely no idea what teacher of the year means, but my friend Dennis (also a teacher) tells me that Toyota has a teacher of the year award, too. There are so many such competitions advertised throughout the country. I am certain that there is a designation for a national teacher of the year; there is the designation of a Wisconsin educator of the year, and at my university there are several such awards in a variety of educational settings. I suppose that across the nation there must be a great many teachers of the year—some of them actually write books after they become teacher of the year decrying the conditions in the public schools in which they became teacher of the year. Perhaps before that nomination they were too busy teaching to have the time to write.

But this particular contest advertisement sponsored by the Disney Company has been clearly constructed with money, which I say might have been better used for an actual classroom. In fact, the classroom constructed for this television scene is not even a classroom: actually, it's a series of rapid mid-range and close-up shots of children engaged in joyous activity in what we *imagine* must be a full-blown classroom but is never shown complete. Even the separate pieces, however, do not look like the pieces of any classroom I have ever seen. The resources available to the classroom here seem limitless and the children oh, so very bright-eyed. I would my daughters look and be so in their daily classrooms. All of the colors on the screen now have nothing to do with the real world colors of the school; indeed, these colors do not even appear to be of the real world. I do not recognize this classroom at all! Here, there are no sniffling children, nor ones who pass gas (that is not the traditional schoolyard word!). Perhaps there is a single child in some classroom who looks like the little girl who marvels at her weight on Jupiter, but I am doubtful if they have her unnatural incandescence. There may be several classrooms where such activities may occasionally take place—but I know of absolutely no classroom that has the means to allocate such funds for even the smallest piece of the resources of the classroom displayed in the commercial. That spot was made not by educators but by media people who are out to associate Disney (without the Walt) with education. It is all so fake. And what of the teacher they will choose? What will be the reward? A trip to Disneyland with his/her class? Or the limitless resources of the infomercial?

I cannot understand what a person would have to do to be nominated teacher of the year. It is easier to imagine what one would have to do to receive the opposite designation: a self-absorption that would make all of the children disappear and turn all of their magical selves back into little mice and overripe pumpkins at the clanging of the school bell. Actually, I have known very few such egos myself, though as a society we prefer mice to children and pumpkins to magical carriages. Most teachers are concerned, diligent, and extremely dedicated to their efforts in the classroom. But we are a society defined by its products, and efforts in the absence of clear products are invisible. Teachers resort to efforts that yield measurable products so that their efforts may be accounted. So many worksheets and so much testing.

And we are a society that depends on external rewards for confirmation: Oscars and Emmys, Grammy and Tony Awards, Opies and Country Music Awards, and Teacher of the Year and Pulitzer Prizes. The Nobel Prize is awarded for a lifetime of achievement! When a person receives one of the show business or literary awards, it is for the stellar accomplishment during the previous year of a product or a single event. The Pulitzer Prizes are for singular productions. They offer the National Book or Film Critics' Award for a single book or cinema production. Player of the year awards in sports events are finally reducible to batting averages or earned run averages, tackles made and/or avoided, passes completed or points scored per game. I mean, it is all so quantifiable. But I wonder how one would measure the accomplishments of a teacher during the previous years: shall it be numbered in books read in the class, or by each student, or perhaps, by the teacher? Shall we measure the chapters covered in a specific text? Or shall the evaluation depend on the ascendancy of test scores? It is not uninteresting to reflect upon what teacher of the year awards might suggest about notions concerning teaching and learning in school. What must happen in those classrooms of award-winning teachers that should make me aspire to emulate them? What should I be doing that I am not already doing (or obviously not yet doing with sufficient energy or productivity)? What have I not yet thought to attempt?

My life as a teacher has been consumed by these questions; much of my academic writing and thinking during the past several years has concerned just these classroom and curriculum issues. This sabbatical is meant as a respite from such work and an opportunity at renewal. What

am I renewing? What should be the result? At the end, shouldn't I at least
be in the running next year for teacher of the year?

✡

I have just had a discussion with my chairperson at *La Reine de Dairie*.
We meet there occasionally to sample the latest flavors and consider our
lives in education and sometimes, as parents. Not two weeks ago—
indeed, not too long after the beginning of my sabbatical leave—we set-
tled there over Blizzards and chocolate sundaes to discuss our children
and education. We met to consider our work in the fields, as it were. I
hoped that this foray was not in violation of sabbatical commands and
especially since while we sipped and guzzled, my chairperson took the
occasion of my sabbatical absence to pointedly express his personal and
professional concerns about *my* work in the fields. Perhaps, he considered, during my sabbatical I could rethink my commitment to my work;
he meant in the present instance merely to offer some guidance and
direction. (I think, in the end, I may have accomplished what he desired
but with an outcome I do not imagine he might appreciate. I have during this celebratory time of my sabbatical indeed, gone far afield, so to
speak). During the course of our discussion, at times energetic and
unquiet, he characterized my efforts in the classroom as the work of a
master craftsman. I did not know if this was meant as a compliment or a
critique. I have regularly expressed (and often to him) that to a large
extent teaching is for me an art; art, I believe, represents an enactment of
a vision with skill and concern. *Of a vision!* That is, the production of
something that exists greater than and beyond the present moment and
which may not ever be fully known or even realized. A vision defies specificity. The work in the classroom becomes art when the artist *cum*
teacher, or the teacher *cum* artist, enacts her educational vision. Art, I
believe, creates the artist even as the artist creates the art.

Some visions, I know, are more acceptable and receive greater display
and respect than others, though I am also aware that these particular preferences exist in historical contexts and often vary with time. Graffiti,
which at one time earned the painter a jail term and/or stiff fine as a public nuisance, later was purchased for huge sums of money as the embod-
iment of a modern art. The social, hierarchically-structured distinction
between art and craft accords greater honor, respect and remuneration to
the former, though the precise discriminations that distinguish them are
rarely so clear. The designation "arts and crafts," an activity in which I
engaged in day camp many summers ago, distinguishes between the two
products and therefore, between the two processes. It may be a specious

distinction! Years ago I had to abandon the distinction held in society between art and craft (art show vs. craft fair, for example), because I learned that some fine crafts (Grecian urns) had since become fine art. I had come to understand this distinction between art and craft to be a qualitative one and as so, subject to historical conditions and therefore, subject to change. Perhaps, because I could not afford the former I attributed its quality to the latter, which I could manage to purchase. I now recognize that even though the two may be to some distinct—indeed, perhaps to have historical distinctions—they are not always distinguishable. I wondered to what my chair referred when he called me a "master craftsman." In this present conversation with him, I was not certain if and how my chairperson was distinguishing between the two types of effort. I did not know what exact topic was being broached. Should I be flattered or offended? What did he think I was doing in the classroom? Indeed, what did I think I was doing there?

I remember some years ago a scandal at the Metropolitan Museum of Art in New York City when a host of paintings attributed to Rembrandt were discovered to be products not of the master but of his studio. A whole host of paintings deemed priceless on Monday were devalued to prices on Tuesday. I wondered what in the paintings had overnight changed—I mean, if someone could imitate a Rembrandt so capably for so many years, then what, indeed, was the difference between great art and great reproductions. Good technicians were all that were required for the production of art; the master craftsman could produce a Rembrandt-like work with extraordinary skill and even genius. If it were true of great art, perhaps the same might be true, then of teachers; teachers fashioned others to serve like them and the master craftsman, a quality ascribed to me, was one who had an exceptional ability of mimicry from some unnamed (and unknown) master and who could masterly pass along the skill. My chair had a point. I might have been flattered. But I wasn't.

Teaching I suspect, partakes of the nature of both art and craft. I think teaching is an art: each class happens once and may not be reproduced; and teaching is also a craft: everyday the classes meet and the teacher must work in them. Teaching is a craft—there are general methods and methodologies that are particular to the classroom; and teaching is an art—every method produces a different product according to the efforts and personalities of the participants and the particular historical conditions then prevalent. I think a teacher must be an artist and a craftsperson; was this included in the chair's designation?

And so, eating my chocolate sundae I listened expectantly as the chair began again to speak. I was a master craftsman, he continued and I had developed my skills and my tools to a priceless degree. I was flattered. But, he bemoaned, my tools were designed for work with walnut, and all I was delivered in my classrooms was pine. Refusing to change my tools or my methods, I had managed to produce some fairly misshapen boxes with my wonderful, but inappropriate tools and methods. I needed, he suggested, a new set of tools and practices. Perhaps I might use this sabbatical as a time to fashion such tools and practices. Or, he suggested, I might seek some new materials . . . elsewhere. I was not flattered.

It was, I recalled, Rembrandt's original vision that made priceless the originals. It was the world Rembrandt had conceived that others had reproduced. Oh, the studio apprentices were technically brilliant—their skill was worth some money—but it was the vision of the master that was, finally, priceless. The artist is the one with vision and the dedication and skill to realize that vision. Rembrandt's studio students and workers reproduced his vision flawlessly. Indeed, some of the studio apprentices might have even surpassed the master in technique!—but it was their vision of Rembrandt's world that they offered. They never aspired to their own vision; they were content to live with someone else's.

Without vision, Isaiah tells us, the people perish.

Or perhaps it was sufficient to the studio workers to (re)produce another's vision as their own. To see the world through the eyes of another. Their work was not inconsiderable nor valueless. It was, after all, to the basement and not to the trashcan that their work was relegated by museum officials. Finally, however, their effort was adjudged not as invaluable as that of the master. It is by his vision we live and not their work. It is his vision that is beyond measurement.

We are not all Rembrandts, but with all of our energies and our skills we might aspire to our visions that we would not perish. Teachers must have visions and must inspire students to envision. Those visions must be of a world greater than that of the classroom and grander than what can be contained in the materials we employ. When we do not have vision, we are condemned to the narrow present and see no purposes to our acts except to fill the time before the bell rings. If we cannot inspire our students to have vision, then we cannot accept our students as Other; rather, we insist that they impersonate us by assuming our vision even as we deny them their power to be character. Now, I do not think we ever completely realize our visions or we would cease to live—without vision the people perish—but teachers must never cease to live and work by their visions. I think it is this effort that ties us to the wonderfully mundane of the classroom. Perhaps our classrooms ought to be spaces where glorious failure is celebrated. Before he died, Rabbi Shelomo Hayyim (Buber,

1947) said to his sons: "You are not to think that your father was a zad-dik (a wise man), a 'rebbe,' a 'good Jew.' But all the same I haven't been a hypocrite. I did try to be a Jew." We should at least aspire not to be hyp-ocrites; we should try to be teachers. We must have original visions and not live by the visions of others. We must be artists that we would make artists. Art often troubles the waters. I told this to my chair.

I also told my chair that his metaphor was quite lovely *if* people were wood, which they are not (though I have, of course, known not a few blockheads). I continued: to be a teacher was to deny the power of that biological determinism and to be committed in teaching to the belief that education would play an enormous, though not complete, role in deter-mining the nature of the developing product. A tree is never free to choose any role but that of the particular tree it is; to our modern sensi-bility it has no consciousness and certainly no free will. But people are not trees and, within limits, do have choice in their identities and in their growth. A human being may choose to do and be many things, and a teacher's enacted visions begin to suggest to students possibilities for that growth and can motivate them to consider the responsibilities attendant upon those choices. A teacher's visions can change the world.

Oh, I do not think that I am succumbing to an ideological idealism here: though the army suggests that by enlisting in the armed forces one could "be all that all you can be," that is true only within severely con-strained limits. I often tell my students that telling the drill sergeant that you would like to be a circus clown will not excuse you from any aspect of basic training. Nor, I tell them, is that career choice possible within the parameters set by the outcomes accepted by Army training. Karl Marx said that men choose their own history but not in circumstances of their own choosing. We will never be completely free—that is why we must be responsible to others—but within limits that include history, we have freedom. Human choice demands that we understand the possibilities for, and consider the consequences of, our acts. Such learning is the work of the classroom. School ought not to be anything like the army or any other organization based in a cold rationality. The ends of the classroom are always beyond it and somewhat out of view; how could these results be so linearly organized in rigid scope and sequences? If indeed, study has replaced the sacrifice as mediator for human measure, and as the teacher serves, willingly or not, knowingly or not, in the position of priest in the exercise of these, as it were, religious acts, then it is in study that we mean by the present to change the future and rewrite the past. And as prayer accompanies the act of sacrifice to ensure its holiness and acknowledge the limits of our humanity, then perhaps study ought to be understood as holy as prayer; in the practice of the sacrifice we aspire again to holiness. In the act of study we renew our dreams. It is the teacher who might represent this idea in the classroom. But first there must be the opportunity for vision and for art.

And so I acknowledged that the tools and methods my chair told me were inappropriate were, indeed, inadequate *if* I aspired to his vision of education. But, I insisted, I didn't. I reminded him that even the best and the worst tools are yet adaptable to new purposes and thus, might serve ends different than those for which they had been designed. I recall walking into a surgical supply store in the early 1970s and asking to purchase a Kelly clamp. One single Kelly clamp. These are delicate and carefully wrought tools used in surgery to tie up bleeding veins. They make surgery possible and save not a few lives. The clerk asked why I had need of a single Kelly clamp. I could not tell her I was going to use it for a roach clip.

However, it worked marvelously.

The artist's tools within certain, albeit broad, limits are specific to the task and not necessarily to the material. One could study the *affichiste* artists of the early part of the twentieth century to learn to what ends materials might be originally put quite apart from their original purpose (Block, 1999). Recall the materials innovatively employed in the collage art of Picasso and Braque. Materials have themselves no intention but must be given purpose by the artist/crafts person. And the skilled craftsman's tools might serve purposes other than those of the craft and yet produce art—hence, the washtub bass fiddle and the musical saw. The value of the tool is in its use, and it is the artist *cum* craftsmen who gives the tool its value. Nor does the material care what other tasks the tool has performed: it does not matter to the tree what music has been played on the saw that cut it down; nor do I think the music knows its relationship to the tree. The value of the tool is not intrinsic to its form and matter, but resides in its use. That use depends on the imagination of the mind and hand, which employs it. Humans are wonderfully pliant and valuable, and indeed the finest tools must be employed to achieve the sacred end of education. But our tools must be variously used and with originality. And we need not legislate requirements and restrictions that determine their uses. We would make new tools out of the material of the old to give power to our students to learn not only how to have vision but to search out for its realization. First, however, the teacher as artist must have vision that he/she would know how tools might be employed. The teacher as craftsperson must have skill to use those tools. And sometimes the finely crafted tools that felled the tree will make music concerning the forests.

I suppose it will be said that a narrow vision—national standards and standardized curriculums—is still a vision, but I would counter that great art presents great vision: it is to the development of individual potential in the context of social responsibility to which the democratic classroom in a democracy must give voice and not to the progress of the individual at the expense of the welfare and care of others. As we learn in the book of Esther, she who saves another saves the world; he who destroys the other destroys the world. Study in this sense must derive from and lead to vision—to the quest for a social justice and not a standardized achievement. It is only after the trees have been cut down and processed that their wood may be standardized for use in certain vocations and/or productions. I would not sacrifice the forest for the trees. Nor do I understand the role of education as equivalent to the processing of, or the working with, wood. It is not graded standards we seek but un-gradable visions. Teachers might enact the belief that the classroom is a production of art and the teacher the artist. To this end, tools might be employed, though not always as intended by their creator. The standardization of curriculum denies the very substance of the teacher's work and the character of the classroom he/she might produce. Standardization demands conformity and contradicts education. Standardization is the production of ceramic coffee mugs and not Grecian urns. Perhaps it is time to reconsider whether our teachers of the year ought to be known as great artists and not skilled technicians. Perhaps it is time for teachers to understand their position in the classroom as artists and not factory workers. Perhaps such consideration might alter the criteria for teacher of the year.

In Malachai, the people complain that in this world God rewards the evil and punishes the good. Of what value, they ask, is good behavior in such a system? It is proposed that in today's classrooms, the rewards (monetary) be given to those who produce the highest scores on standardized achievement tests and that those whose scores are not sufficient, despite the conditions in which they work, be punished. Many good teachers despair. Malachai's answer offers some hope, I think, to the change in educational direction I here propose: Malachai responds that though in this world it *appears* that the values are wayward and injustice prevails, and even though at present this may be so, *yet the day will come. . . .* I wait yet.

I did not leave the conversation with my chair either satisfied with myself or convincing to him. Is the function of the teacher the achievement of academic standards or the realization of individual growth or an achievement of social justice? The issue appears regularly in the daily papers: I see in the newspapers that at least ten states will link teacher salaries to academic results. And what will they measure—how the student measures the flagpole from the length of its shadow, or how the student is measured by the depth of his/her responsibility? It is clear today that the best teachers are considered those who will achieve for their students the best test scores and the best educational methods are those that will attain higher scores on standardized tests. The superintendent of schools in Menomonie, Wisconsin recently sent around a notice prior to the beginning of the annual standardized testing in the schools. He stated proudly that though our fourth graders score well in the state tests, our eighth graders do not so well succeed. He promised that our curriculum would be better adjusted to the standards set by the tests so that the scores might be improved.

When I was in school such behavior was referred to as cheating. Education should, I think, serve higher ends.

✡

I would like to think about teaching and learning using a set of discourses not commonly—indeed, not at all—employed in pedagogical conversations concerning the American public school. I read in Talmud . . .

These particular Talmudic stories pertaining to teaching and learning about which I would like now to think occur in the midst of an intricate discussion concerning labor relationships. These stories of teaching and learning that I intend to here address occur as a digression from the discussion of the halakhah (law) concerning obligations for which an employer is responsible to her/his employees. Perhaps it is not far afield to consider my efforts in any classroom as work: teaching entails relationships, which are laborious. I never think of myself as not a worker. Somewhere, there is a contract to which I am subject setting terms and conditions of employment. I receive—and have always received—a state-issued paycheck with all the requisite deductions deducted; I pay the requisite tithes. I am daily obligated to certain times and to certain responsibilities that must be performed to certain standards that are particularly set by the institution by which I am employed. In these certain times, these standards multiply like fruit flies and annoy like gnats. But to assure the maintenance of my position, I, at least cursorily, do obeisance to the requisite mores. I attend department and other academic meetings. There is so much in my school life that reminds me that I am an employed worker. This sabbatical leave is supposed to be a Sabbath from that work. I do not go to classes nor to meetings. I do receive the same monthly check, though, and, as I have said, I wonder now what I should be doing to deserve it.

Certainly, my students always refer to their efforts as work—and I do not think there is much hint of pleasure in their tone. They complain, "This class requires too much work." They say, "I'm working really hard for this class." Antonio Gramsci (1971, 42) reminds me that "Many people have to be persuaded that studying too is a job, and a very tiring one, with its own particular apprenticeship—involving muscles and nerves as well as intellect. It is a process of adaptation, a habit acquired with effort, tedium and even suffering." Perhaps it is to the students he should be talking.

I have myself often experienced periods of tedium and even suffering in my study. I have learned to labor continuously and appreciate contingency. For example, despite his supposed brilliance, I have never appreciated Wordsworth's *The Prelude*. Whenever it was assigned, I would intellectually groan. But I almost always completed the assignment, perhaps more often barking at a few words than biting down upon them. I recall absolutely nothing from the experience except this barking. Yet, though I feel even now no diminishment from this failing and though I chose not to study this particular Romantic person/event in literary and cultural history, I recognize in my intellectual architecture a certain necessary but rickety bridge across which I may sally when I am working with some other matter upon which Wordsworth might shed light. I mean, I did remember, as you will discover below, Wordsworth's definition of poetry! This Romantic poetry is not a burned bridge; rather, it yet connects, albeit tenuously, my present to my past. If necessary, I can cross that bridge when I come to it; I just can't always tell what might lead me to that bridge, or what I might find on the other side. For example, I had no idea I was going to bring William Wordsworth into this present writing, and suddenly he appears.

I have sat with brilliant texts in horribly unpleasant contexts. The texts, of course, always survived the experience, though sometimes I did not know it. I was driven to tears by the context in which Dickens's *A Tale of Two Cities* was first taught. But, when later I had made the study and learning of literature my life and work, and had even been to Paris, I was thoroughly captivated by the text in which I again traveled. There are many such bridges in my intellectual geography—tenuous and fragile connections that can, when necessary, support vague wanderings. Thoreau says somewhere that many men fish all of their lives without knowing that it is not fish they are after. I have always woven narrative to produce a contiguous world. I would teach such weaving in my classroom. I have never had an aversion to learning, even though it has always been difficult. I would offer such comfort in the classroom. Perhaps there is a difference between studying and learning, the former subject to tedium and suffering and the latter to . . . well, to what exactly? When I experience learning I feel satisfied, as when I dine amidst fine company engaging in conversation and where the food is more than passable. I love to learn—somehow the activity of learning evokes a character who is satisfied and content, and the pleasure I experience is not unlike that of the pleasure of sex. There is a considerably physical aspect to the act of learning—physical and mental activity, tensions and negotiations, release; perhaps these connections are what account for some of the sexual misbehaviors in the relationships between teachers and students. When this writing goes well I feel powerful, and when I feel powerful I sense my whole body, and when I sense my whole body I think of sex. Sex is often a matter of playing with control, and good learning is, too, the exercise of letting go and holding on. It is not uninteresting that the Talmud on which I wish to think also makes connections between study and sex. But I am, I think, far afield for now. Which is a Talmudic characteristic.

Sometimes when I feel powerful I go out running.

If I am on sabbatical and I am to study in a way I have not been able to study before because of the obligations of "work," should I be prepared to experience tedium and even suffering? I know my students experience such states: they inform me so regularly. If I am on sabbatical and I am to rejuvenate myself, how does one arise refresh from the experience of tedium and suffering? Of course, to continue the analogy earlier begun, sex is sometimes wonderful and sometimes tedious.

CHAPTER 4

THE FOURTH SAPLING

I have been studying Talmud down here in my basement for the past several years. It has at times been both a maddening and a joyous experience. Moshe Halbertal (1997, 1) admits that he had been taught that hell was not where the wicked were consumed by fire, but rather, where they had been made to study Talmud. Heaven, on the other hand, was the study of Talmud by the pious. Often, in my study, I have pondered into which category I might fit. The Talmud is a book of exasperating complexity whose full depths I may never plumb. Talmudic scholar Adin Steinsaltz (1976, 4) describes the work as

> the repository of thousands of years of Jewish wisdom, and the oral law, which is as ancient and significant as the written law (the Torah) . . . a conglomerate of law, legend and philosophy, a blend of unique logic and shrewd pragmatism, of history and science, anecdotes and humor. It is a collection of paradoxes: its framework is orderly and logical, every word and term subjected to meticulous editing, completed centuries after the actual work of composition came to an end; yet it is still based on free association, on a harnessing together of diverse ideas reminiscent of the modern stream-of-consciousness novel . . . The Talmud treats abstract and totally unrealistic problems in the same manner in which it refers to the most prosaic facts of everyday life.

The Talmud is the central text of Rabbinic Judaism which, at times, has replaced even Torah as the primary canonical text. It is in essence (but not essentially!) a commentary on the Mishnah, itself the redaction of centuries of transmitted oral law finally compiled in about 200 C.E. by Judah Ha-Nasi. Ha-Nasi, which means the prince, appears throughout the Talmud referred to usually as "Rabbi" or "Master." The Mishnah consists of rulings that deal with all aspects of the details of daily life as they evolved over the centuries of Jewish history and exegetical analysis of Torah.

Now the Talmud is a redacted document which consists of the Mishnah, narratives enunciating law, and the discussions in which the Rabbis engaged about that law and other related (and unrelated topics)—the Gemara. Talmud means study. Problematically, the conversations that are recorded appear all too often in abbreviated and incomplete form. Furthermore, there are dramatic and remarkable digressions that often lead the Rabbis and their readers perplexingly far from any path. At the very least, there is in Talmud very vague pronominal references that would have driven mad Miss Bueschel, my ninth grade English teacher. Talmud is not itself merely notes per se, but narrative constructed from such notes. I hope my students write such cryptic and rich notes. I have spent my life in texts and have seen none that approach the style and substance of Talmud.

The editing of the Talmud derived in large part (as the *sugyot*[1] that I will study in this text suggests) from the work of a particular scholar, Rav Ashi, who constructed the overall structure of Talmud based on the minutes of the discussions that took place at his academy in Babylonia, which served as the most important institution of Jewish learning at the end of the fourth century. Over the years these discussions inspired further exegesis, by Rashi and other French rabbis among others, which then became part of the discussion. Commentary from later periods of study and cross-references to other sections of Mishnah or Torah also appear in the margins of the pages of Talmud. To look at a page of Talmud is to attend to and participate in a wondrous conversation that encompasses many centuries and a variety of texts and that takes place over hundreds of centuries.

My life has been spent in the academic study of texts. I have heard often the mermaids singing.

The Talmud, Adin Steinsaltz says, is a kind of record of discussions in the academies. Talmud is a canonical text that sets as its task the exposition of Torah, one of the supports on which the world is founded. Simeon the Just, who lived in the time before the destruction of the Temple (70 C.E.), said that "Upon three things the world exists: upon the Torah, upon the temple service, and upon the practice of charity" (*Pirke Avot*, 2002,1, 2). With the loss of the Temple, one of the three pillars, as it were, has been removed; without it the sacrifice could no longer be carried out. The Rabbis replaced the practice of the sacrifice with prayer and continuous study. Indeed, the whole first tractate of Talmud Berachot, concerns the subject of prayer. The practice of prayer becomes in Talmud the subject of interpretation and halachic (legal) findings. There we read that even God puts on tefillin—the phylacteries that are commanded to be worn as accompaniment to prayer. According to Talmud, even God engages in prayer.

With the loss of the Temple, the ritual exercises of the sacrifice were transferred to the ritual practices of daily life. What the priesthood in the Temple had been designed to mediate through the practices of the sacrifice could now be effected in the home and by the people in general: the Jews were to become a nation of priests. The purpose of study was to discover what Torah required in the exercise of daily life. The canonical texts contained the blueprints for a righteous life, but those texts first required study and interpretation. For example, Deuteronomy 1:12–13 prescribes the appointment of wise judges because Moses cannot "alone carry your contentiousness, your burdens, and your quarrels." But the Rabbis wondered about the constitution of a law court. How many judges are required to sit in judgment to ensure justice? Since all authority derives from canonical text, then it is to the canon that the Rabbis go to establish the law court. And the Rabbis of the Talmud discover the remarkable answer to their query from Psalm 82. There they read, "God stands in the Divine Assembly, in the midst of judges shall He judge." From this the Rabbis argue that only in the presence of a minimum of three judges (in the *midst* of judges shall God judge) is God's presence assured.

Now the Rabbis know that God does not participate in instances of justice that are intended simply to make peace—humans are free to act. Peacemaking is a behavioral standard measured by God's words and not by God's presence; that is, for such judgment, God need not be present. And so the Rabbis wonder, what would define the justice that ensures God's presence in the midst of three judges. That is, what is *more than* "peacemaking." Or conversely, what is *merely* peacemaking and not seeking justice? The same Psalm 82 argues that this established court will "dispense justice for the needy and the orphan; vindicate the poor and impoverished. Rescue the needy and destitute and deliver them from the hand of the wicked." Thus, the Rabbis reason, any other judicial matters than these are considered merely making peace and would not demand God's presence. The presence of God requires issues of righteousness and justice. The attainment of holiness that requires justice and righteousness is thus fixed by textual interpretation.

Similarly, since Psalm 82 says that God stands in the Divine assembly, the Rabbis wonder under what other conditions God's presence is ensured. "How do you know that if ten people pray together the Divine Presence is with them?" Since it has already been established that a congregation consists of not less than ten, then the Rabbis argue that since Psalm 82 begins with the phrase, "God stands in the congregation of God," then the Divine Presence enters the *shul*[2] when at least ten are in prayer. Thus we might conclude that the Rabbis hold that true justice *and* true prayer invite God's presence. Prayer and study are not unrelated.

But, the Rabbis wonder, what about study? Is it, too, holy enough to invite God's presence? The Rabbis ask, "And how do you know that if two are sitting and studying the Torah together the Divine Presence is with them?" For their purposes this time, the Rabbis turn to Malachai 3:16. In this passage, God remonstrates the people, Israel, for turning from God's ways, and the people complain that there has been for them no benefit from serving God; we have continued to suffer, they groan, and the evil have continued to prosper. Malachai responds, "Then those who fear God spoke to one another, and God listened and heard, and a book of remembrance was written before God for those who fear God and those who give thought to his name." To have *thought* upon God's name means to desire to fulfill a commandment but to be prevented from immediately doing so either by force, accident, or circumstance. The Rabbis' interpretation suggests that study is action intended to lead towards holiness. Thus, the Rabbis reason, that when two are studying Torah, speaking of God to one another, God is present. Studying together is holy work.

Which is all well and good for a pair of students, but what if only one is studying by his/herself, as I am quietly and solitarily doing down here in the basement? The Rabbis wondered if singular study as well as paired study invites the Divine presence. And they acknowledge that Exodus 20:21 authorizes individual study; there God says, "Wherever I permit My name to be mentioned I shall come to you and bless you." Even as I have learned that the pursuit of true justice—concern and care for the Other who is always in need of me—enables God's presence, so, too, the practice of true study takes place in the presence of God and pursues righteousness (Berakhot, 6a). A man studying alone invites God into the study because that man's study is founded on the necessary execution of justice for the needy and the orphan, the vindication of the poor and impoverished, the rescue of the needy and the destitute, and their deliverance from their oppressors. Study achieves righteousness in action. I note that for the

Rabbis it is the *practice* of study and the *practices* that derive from it—not
its assessment—that makes study holy.

I note with consternation that our educational leaders and government officials take the opposite position. For them all value of study
resides in the assessment. Student progress will be adjudged through test
scores and not in deed. The purpose of study for the Rabbis is for the
benefit of the needy other. On what basis is assessment made in our contemporary times? Where might we discover the holiness in the perfect
report card or a high score on the standardized test?

Study, Talmud says, is greater than the performance of the sacrifice, a
brilliant strategy to transform the cultic practices and priestly hierarchy
of the now destroyed Temple into the daily exercise of life. I have been
studying all of my life and for the past twenty-five years have been a veg-
etarian. If the sacrifice mediated between the Divine and the human,
then perhaps I can consider that I have been engaged in holy activity,
though I here confess that too often I feel that in study I have irrepara-
bly sinned. I might think on that! When I have studied and when I have
taught, I too often turned my mind to purposes other than justice, like
the achievement of standards and set curricula and reputation. I have had
to fill so many grade books. I have assigned so many failing grades. I have
disappointed so many students. I might think on that! I teach people
who want to be teachers. I might think on that!

I have been concerned during this sabbatical with what I am daily to
do while giving my fields a Sabbath rest while yet residing in the midst
of those fields. I had an ambitious plan, but I seem to have above
explained that plan away as a potential misuse of the fields during the
sabbatical. But I cannot get away from study or from the idea of teaching and scholarship and learning. As I have said, study achieves holiness
and seeks righteousness. As I have said, I am a teacher and a student.
Besides, my children are in school. Perhaps . . .

In my reading in *Bava Metzia* I have come across a section that contains some strange stories about teaching and learning. During this sabbatical, while the fields are celebrating a Sabbath for the Lord, when I will
not go out to work in them nor will they oblige my care, I want to reflect
upon my life that has been spent in those very fields. I cannot think about
my life—cannot even consider myself as my children's father—outside of
the context of the school and outside of the processes of teaching and
learning and scholarship. But as a result of my own educational experience, I also cannot think about education outside of the Western philosophical tradition. And so, as I had planned during this sabbatical to learn
a new set of discourses with which I might think about education and

about teaching and learning and scholarship—the life to which I have committed myself and in which my children now learn to practice and grow, so I will remain immersed in the complexities of Talmud.

The Talmud represents a non-Western system of thought. Two of my teachers, Emmanuel Levinas and Abraham Joshua Heschel, teach me that it is in the Talmud that I might discover an alternative to Western logic and philosophic systems. My research agenda focused on introducing into the Western curriculum discourses Jewish—Talmudic, Rabbinic, Hasidic, and Kabbalistic—ways of thought by interpreting the work of Joseph Schwab, a Jewish scholar engaged in education (Block, 2004). But I have first to actually learn those discourses myself and use them and reflect upon them. My sabbatical leave is a time to reflect upon my many lives in school, and I can find in Talmud a different lens than any I have previously been given and previously employed. I want also to think about the whole of my life and about my two young children who go to school—but I cannot do so outside of the context of my life in the fields—that is, without considering issues of teaching and learning. I have spent my entire life in school. This sabbatical affords me the luxury of luxuriating in time and perhaps, producing a poem. William Wordsworth (ah, there he is again!) describes a poem as ". . . emotion recollected in tranquility." Could there really be such a thing? Isn't tranquility an emotion that would color the recollection? I have experienced some powerful emotions in the classroom. I want the schools my children attend to be exemplary; I want my children's teachers to be wonderful. I am a teacher educator. If I am only for myself, what am I?

During each semester I tell students at the University that my class is notably rigorous. I justify the design by explaining that my students must become excellent teachers, for there is the chance that one of them will be my child's teacher, and if they be not my child's teacher, then certainly they could become the teacher of someone else's child—for this reason they must aspire to distinction. Inevitably, one student turns to another and whispers, "I wouldn't want his child in my class." I acknowledge that request silently.

I wonder now within these pages if anything I could teach would make anyone a better teacher. What am I to teach?

NOTES

1. A *sugya* is a self-contained deliberation in the Talmud.
2. I use here the Hebrew term for house of prayer. *Synagogue* is a Greek word.

PART II

ON ENTERING THE FIELDS

CHAPTER 5

THE FIFTH SAPLING

The Mishnah (82a) says,

> If someone hired workers and told them to come early or to stay late, [in] a place where it is customary not to come early and not to stay late, he is not permitted to force them. [In] a place where it is customary to feed [them], he must feed [them], to provide [them] with a dessert, he must provide [it], everything in accordance with local custom.
>
> It once happened that Rabbi Yohanan ben Matya said to his son: "Go out [and] hire workers for us." He went and promised them food. But when he came to his father, he said to him: "My son, even if you make [a meal] for them like Solomon's feast in his time, you have not fulfilled your obligation towards them, for they are the children of Abraham, Isaac, and Jacob. Rather, before they begin work, go out and say to them: 'On condition that you have no [claim] on me except for bread and beans alone.'" Rabban Shimon ben Gamliel says, "He did not need to say [this]; everything is in accordance with local custom."

These are laws that regulate labor relations. And the halakhah, the law, the way one should act, or literally, "the way in which one goes," refers here to the power of local custom to restrict what an employer can require of his workers. Unless something else has been previously stipulated, all conditions of employment—from responsibilities to benefits—are contingent on practices established in and by local custom. This conflict between local custom and national standard has beset education in the United States from its public beginnings. It is an argument between those who advocate for local control and those who require more bureaucratic structures. We are certainly not at the end of this debate; indeed, the rhetoric from both sides increases daily in volume and acrimony. Who should be responsible for the organization and running of the school system? Should what the children in any community study and learn be set by the wishes of the local constituency, which includes the parents, or should it be set by the mandates of the state and national government and professional educators? Who should be responsible for the standards by which teachers are accredited and curriculum developed, the people whose children are subject to both, or the state, which at least theoretically advocates an equality and excellence in educational opportunity that requires standardization? What role should federal government play in the regulation of the local and culturally-diverse public schools? Of course, if all is organized and practiced according to the policies of local custom, then what chance is there ever for change and innovation?

Perhaps that is one conclusion that may be made from Rabban Shimon ben Gamaliel's statement. Rabban Shimon declared that the exact designation of food promised to the laborers in the story of Rabbi Yohanan ben Matya and his son is unnecessary because if "everything is in accordance with local custom," then the meal that must be provided the workers is already set by local practice. Neither son nor father need stipulate the restriction on the meal; no other standard but local custom may be employed. But perhaps Rabbi Shimon's objection of irrelevancy indicates that more is at issue here concerning the power of local custom in the organization of at least these labor relations concerning meals.

In the Gemara—the extensive and, as we will see, circuitous Rabbinic discussion concerning the Mishnah—the nature and identity of local custom is contested. For example, the Rabbis wonder, does the "local custom" that must be followed refer to that of the location where the work will occur, or to the place from which the employees originate? When workers are hired to labor outside their own environs, to whose customs are they obliged? When wages are paid, whose minimum wage standard is to be followed? I wonder now how might such a discussion about this problem affect our current immigration policies and our social attitude towards diversity? The Gemara acknowledges the necessity of work: six days a week shalt thou work, the Torah commands, but the Gemara will wonder, and I must soon consider, exactly how long is a working day? The struggle for the eight-hour day might have been less contentious and violent, perhaps, had Talmud been consulted as primary source material. It would seem that if everything is set in accordance with local custom, then the question of any condition of labor ought to be already affirmed.

An employer has only to learn the local custom before negotiating for workers. Negotiations, in fact, are deemed irrelevant as long as local custom is observed.

But perhaps, there is the rub. For to know the local custom is to understand the entire context in which work must take place and precludes the arbitrary establishment of practices that treat workers as anything less than free individuals. Workers enter into work relationships with certain rights that derive from the histories that comprise their daily lives. To have to know the local custom is to require an appreciation for the integrity of community and social relations, which include their particular history. I wonder how might such understanding inform the conversations affecting multicultural and special education? But the adherence to the rule of local custom prevents anyone from enforcing regulations without regard to long-standing local conditions and traditions and by such obliviousness causing the exploitation of the less powerful workers. Corporate patriarchy is here prohibited by the Rabbis. Here no company stores are permitted. Teacher-directed classrooms, too, are forbidden and, it would seem, course syllabi must forever be negotiated.

But the conversation has only begun. If all is according to local custom, then what so perturbed Rabbi ben Matya regarding his son's actions? Couldn't Rabbi ben Matya just rely on pre-established local custom?

✡

This is obvious!

No. It is necessary where he added to their wages. You might have said [that] he can say to them: "That I added to your wages [was] on condition that you would come early and stay late with me." Therefore it informs us that they can say to him: "That you added to our [wages was] on condition that we would do [especially] good work for you.

Resh Lakish said: A worker, when he enters [town, enters] on his own [time]: when he leaves, [he leaves on the time] of the employer, as it is said: "The sun rises, they gather themselves together, and lay themselves down in their dens. Man goes out to his work and to his labor until the evening." (Psalm 104: 22–23).

But let us see what their custom is!

In a new town.

Then let us see from where they have come.

Where they were gathered [from different places.]

If you wish, say: "Where he said to them: 'You are hired by me as workers according to Torah law.'"

The Mishnah does not mean to question the power of local custom; that ascendancy seems to be already assumed. It is obvious! If everything is according to local custom, then as Gamaliel says, ben Matya need not have been concerned. This is obvious! But the Gemara answers, the stipulation *is* necessary even when local custom is taken into account because every contingency must be accounted for in the labor agreement. The Rabbis argue that it is necessary to explicitly define each condition of employment to ensure that no practice, even one that appears to benefit certain workers, extends practices beyond local custom and results—even in the absence of intention—in inequities. The Rabbis seem to guard against all forms of exploitation. For example, the Rabbis offer, if an employer contracts with some workers to be paid additional salaries and then asks those employees to work extended hours as additional service for that extra pay, the Rabbis state that the workers have the power to deny that the money was given as overtime pay and to assert instead that the monies were intended as remuneration for the especially good work for which they were especially hired. In the absence of further qualification, all remains according to local custom and workday hours may not be arbitrarily extended. That is, the Gemara denies the employer's privilege to use his power to circumvent local custom (daily life) for his own purposes, *even if he is prepared to pay for that change.*

But the issue of the length of the working day has been raised and must be, therefore, addressed. How long, indeed, is the working day? According to Psalm 104:22–23, the worker leaves his home after the sun rises and is, therefore, considered already at work—travel time is included in the workday hours. The worker is then obliged to work only until the evening. But, the Rabbis ask, wouldn't local custom regarding the length of the workday hold sway? Lakish's ruling must therefore concern hours to be set in a new town. Well then, the Rabbis argue, in that case we should simply check the local custom in the town from which the workers derive. But, the other sages object, what if the workers come from different locations where the customs are necessarily diverse?

What the Rabbis are here concerned with is the prevention of outright oppressive labor practices: everything may be according to local custom, but local custom only begins where Torah law leaves off. In the absence of local custom, Torah is the ultimate authority. For example, Torah requires caution that "the sun shall not set on a worker but that his pay is remitted (Deuteronomy 24:15); Torah states: "a worker's wage shall not remain with you overnight until morning" (Leviticus 19:13). Nothing in local custom may thwart Torah law; local custom may be authoritative, but it may not circumvent Torah. This is the argument that Resh Lakish makes above regarding the length of the working day. Lakish argues that in the absence of local custom, Torah law must be applied. Or, for my purposes here, it might be argued that Lakish claims that a labor arrangement in the absence of greater specificity assumes that "I am hiring you as workers whose terms of employment follow the guidelines set by the Torah." Local custom must be always coincident with Torah, but in the absence of clear, local custom Torah is authoritative. That is, it is authoritative once Torah has been subject to interpretation. Lakish argues that terms of employment set by the Torah regarding the length of the work day supersede even local custom, though local custom that does not contradict Torah law should suffice to resolve most labor issues. Then why, I wonder, is it necessary to discuss the length of days or the meals to be provided, as does this Mishnah? What is so disturbing to Rabbi ben Matya?

It is from this Mishnah concerning the power of local custom that the conversation I wish to join begins. I can certainly consider the implications of my work from the perspective of this piece of labor relations and local custom. Indeed, I should so consider it because I am not a few years away from retirement, and I have a relationship to local custom that must be considered. In the execution of my labor, I prepare syllabi, assignments, and tests. I assign texts that I have chosen—even students' free choices of reading material are subject to my approval. The local custom I follow is, perhaps, not local: rather, it is ideological. It derives from what I think a teacher should do—I have learned this somewhere. My practice assumes a student I have invented and who may actually be me; I have learned these forms somewhere and they may have nothing to do with my students, finally, do they? Perhaps there is nothing local but my location, and I could be anywhere. And besides, my daughters are in school—is there a local custom that organizes their day and their practice? This Talmudic conversation on the importance of local custom suddenly speaks to my situation in the classroom.

Resh Lakish has said, "A worker, when he enters [town, enters] on his own [time]; when he leaves, [he leaves on the time] of the employer." The question: by what standard is the length of the workday to be measured? Does a worker's day begin when he leaves his house or when he enters the field? Does the day end when the worker leaves the field, or when she arrives home? These are not inconsiderable issues—the employer must pay wages (according to local custom!) based in the actual length of the workday. What work an employer can expect the workers to complete depends to a large extent on the time his workers will be employed in the day. How long school lasts often is fixed by the quantities of material that must be accomplished. In this Talmudic *sugya*, the rationale for the length of the working day (which is not finally agreed upon—its length will depend finally on interpretation and local custom) derives from holy text—Psalms 104:22–23. There it says, "The sun rises, they gather themselves together, and lay themselves down in their dens. Man goes out to his work to his labor until the evening." Resh Lakish, one of the Talmudic doctors, says that the workday begins *after* the animals, who forage at night go back into their dens: the workday begins after the sun rises. However, the worker stays at work *until* the evening—that is, until after nightfall. Except, I suppose, on Shabbat,[1] which begins at nightfall.

Every afternoon Emma comes off the school bus swinging a backpack stuffed with assorted matters—her lunch bag with assorted leftovers remaining; spare shoes; snow pants (which I have since learned fourth graders do not wear despite the snow and the wet and the cold because it is not cool to wear them). I ask her: "How was your day?" "Good," she responds. "Did you ask any interesting questions today?" "No." "What did you do at recess?" "Stuff." It's a great conversation, I think, as we climb our driveway. Finally I ask, "Any homework?" She responds, "Ugghhh, tons!" I wonder now as I consider this Talmudic discussion concerning the length of the workday, whether the school day should end at some time. The Rabbis in this Mishnah say that there is a fixed time for the workday; and so I now wonder, for students, when exactly does school end? There is no question that the Rabbis prioritize study, but what is the relationship between work, study, and the homework schools assign? Might school be considered "work"? There are so many worksheets—whose local custom do they follow? What is the relationship between engagement in study and the completion of worksheets? When does Emma's school day end?

✡

But why must the Rabbis engage in this discussion at all if local custom is to be followed? The Rabbis note that the other mention of the workday in Torah is in Nehemiah (4:15) where its length is measured from "the rising of the dawn until the emergence of the stars." That seems to me an inordinately long day, beginning before sunrise and extending to starlight. But, we are told, the people described in this passage in Nehemiah were attempting to build the Temple Walls, and the work had to go up quickly because the Jewish people were endangered by such enemies as Sanballat and Tobiah and the Arams and the Ammonites and the Ashdodites. Thus, although this passage seems to contradict the earlier one, the Rabbis agree that Torah law defines the workday as Lakish had stated; thus, the workday is defined from sunrise until evening. But when exactly is evening? Psalm 104 concerns God's work—the work God has engaged in to create and to sustain the world: "He who roofs His upper chamber with water . . . He established the earth upon its foundations that it not falter forever and ever." Verse 20 of this Psalm declares that "You make darkness and it is night, in which every forest beast stirs," indicating that night is when beasts roam and forage; at the rising of the sun, however, the animals go into their dens and man comes out to work and labor until evening. Evening is before night. Thus is the working day established.

I wonder how do we justify our school day? When is Emma's evening? There is much talk these days about the lengths of the school day, the school week, and the school year. I know that regularly someone somewhere mentions how children somewhere else go to school for greater periods of time and for a greater number of days and achieve better test scores. School meets elsewhere six days a week. School elsewhere is in session on Saturdays! Of course, I believe such practice argues that Saturday is a legitimate day for school and work, yet we are Jewish and thus, on Saturdays, Emma would stay home to celebrate Shabbat with her family. Tom Keating (1999) has written a book showing how the town of Decatur, Georgia established "Saturday School" as a way to keep Jews from settling in the district.

Lakish's reasoning concerning the length of the workday is opposed by an anonymous rabbi. Often in the Talmud there is no attribution of names, but simply an anonymous statement of a question. In the Steinsaltz commentary, the editor attributes this objection to a group. He writes: "The Gemara objects to Resh Lakish's ruling." I am intrigued by the choice for anonymity. In this case, it seems to have the force of number. "They" argue that we cannot rely on the substance of that verse in Psalms that establishes the length of the working day because the whole thrust of the Mishnah under discussion is to establish the authority of local custom. If local custom is to be followed, then the interpretation of Psalms as setting the worker's day is irrelevant. But the Rabbis are not inclined to dismiss any opinion outright, and so a way is sought not to disprove Lakish, but to make everyone's voice legitimate. Well, someone offers, perhaps Lakish is referring to a specific case where local custom has not yet been achieved? Say, in a new community! Well even so, the response is made, we must then check on the local custom of the worker's place of origin; perhaps they come from very far away. But, it is objected, what if the workers were gathered from different places?

✡

And so, despite the overt references to the length of the workday and to the meals to be provided, I don't believe that the hours of employment or the content of the meal is actually what this Mishnah is about. I mean, these practices are apparently and halachically set by local custom. I think Gamaliel's objection is valid: Rabbi ben Matya need not qualify the parameters of the meal that he must provide his workers because everything is in accordance with local custom. The ensuing discussion indicates to me that what the Talmudic doctors are in fact concerned with here is the necessity for face-to-face meetings between parties in at least work relationships. Mishnah insists, and the Gemara reinforces, that in labor negotiations all conditions must be handled explicitly at the outset of the agreement and that all subsequent doubt be resolved if not in favor of the employee, at least not to her detriment. Thus, perhaps Rabbi Matya should have been precise with regard to the meal to be provided and could not rely on local custom because the exigencies of daily existence make the varieties of local custom illimitable. Torah insists we must remain absolutely just to all of our workers. Rabbi Matya's consternation reflects his awareness of the inflections of local custom and the obligations he assumes regarding them. In the absence of explicit definition, the tradition obliges the employer to treat his workers as the heirs of the patriarchs Abraham, Isaac, and Jacob to whom all Judaism looks as its foundation. What a local custom is there!

Teaching, I think, requires a face-to-face commitment. What is the local custom? What is the local custom we follow? The Talmud suggests here that I must *not* be the one who knows; rather, I must stand mute waiting to be commanded. In Buber's *Tales of the Hasidim* (1947), I find this tale that speaks to my concerns. The Rabbi of Kobryn told this story of himself:

> When I was young I once spent Purim with my teacher Rabbi Mordecai of Lekhovitz. In the middle of the meal he cried, "Today is the day of gifts, the hour for giving has come. Whoever reaches out his hand, will get from me whatever strength in the service of God he desires for himself." His disciples asked for a variety of spiritual gifts. Each got what he wanted and kept it.
>
> Finally the rabbi inquired: "Well, Moshe, and what do you want?" I fought down my shyness and replied: "I don't want any gratuitous gift. I want to be a common soldier and serve until I deserve what I get."

I am a teacher and the students enter the classroom without need of my permission. I work until I deserve them. They are, after all, Other. I must learn by observation of and harkening to the local custom. To look into the face of the Other means that I am not independent and solitary and that my behavior may not be originally established for my benefit. Isn't that what Esther's actions were all about: If the Jews perish, she acknowledges, then I shall perish. The priority remains the welfare of the Other; in that emphasis, Esther comes to be Esther. Perhaps in our overcrowded classrooms and lecture halls, we tend to lose sight of the faces that should command us. We hide ourselves—our own faces—behind the comfort of the lectern in order, perhaps, to avoid these face-to-face commitments so that we might in confidence continue in comfort what we, despite local custom, continue to do. If, as Levinas says, violence is to act as if we were the only one present, then many of us in our classrooms practice violence. This is what I understand Levinas (1990, 8) to mean in "Ethics and the Spirit" when he says:

> I do not only know something. I am also part of society . . . Speaking institutes the moral relationship of equality and consequently recognizes justice. Even when one speaks to a slave one speaks to an equal. What one says, the content communicated, is possible only thanks to this face-to-face relationship in which the Other counts as an interlocutor prior even to being known.

To speak to another is to hear that other's voice in the form of a question or demand addressed to me. To be part of society is to move outside my self. Everything over which I assert control, by definition, must belong to me—is part of me—and cannot, therefore, be free. To speak to another, says Levinas, must acknowledge her freedom, but to claim to possess anything—even knowledge and knowledge of another—is to commit a violence. To claim to know another person is to transform that subject into an object that can be then contained. This, I learn, would be the practice of violence. I am the one in the classroom who knows, but if I claim to be so I have already instituted the practice of violence against my students. The face of the Other to which I look is the acknowledgment of the Other's claim on me: the face is what I cannot make an object because that would be to possess it. I can give a failing grade to a thing but never to a face. And vice versa: the ascription "that asshole teacher" can be applied only to a thing.

I wonder if there is enough time in the semester to stand face to face with my students. Thank goodness. How would I have time for anything else?

I think that it is to this imperative of face-to-face meeting to which the Mishnah above refers. There is no end to this obligation. Thank goodness. How else begin to fulfill my responsibilities to students?

And the local custom to which we are so heavily responsible?—It must be elaborated face to face. Was that a bread made of beans that I have been promised for my meal, or have I been offered bread and beans?

I am a worker. I have workers. Of course, I am called professor and I call them students. I was once a student to teachers who had been educated by teachers to be teachers. I am a teacher to those who are learning to be teachers. I am a student to those who would be teachers. The moral responsibility that falls upon me rests in the face of the students. "For Judaism, the world becomes intelligible before a human face . . ." Levinas writes. Not in their gazes, which I might possess, but in their faces, which I might never know, and which command me infinitely to act, rests my professional responsibility. Rests my fatherhood.

✡

I read in the faculty handbook that teaching, research, and service are the duties of the faculty; but what the local custom is with regard to these activities is left undefined. Is it the local custom from where I derive or in which now I work? And what of the workers who gather from different places? What local custom obliges them? As the Mishnah is the law and the Gemara is the discussion and disputation concerning it, then perhaps the faculty handbook requires a Gemara—a discussion and clarification of the meaning of the law. Who should write it? The same would be necessary regarding the relations between student and faculty. Or in that absence should our authority be finally Torah? We would all be infinitely bound to the service of the Other! Maybe the local custom to which I acceded was not really local but localized—to the discourses of *this* University and *these* classrooms. These customs have become so customary that they are now taken for granted. Perhaps that is a problem: perhaps dissatisfaction arises because explicit terms are not regularly proffered to the worker but are, rather, assumed by those already in authority: without that explicitness we all feel due the meal appropriate for the children of Abraham, Isaac, and Jacob. That meal, *proffered unsolicited and complete*, the Gemara teaches in wonderful detail, is the exemplary meal. Each and every word of the story of Abraham's generosity to the visitation of the three angels defines an ethics, which demands absolute consideration of the needs of the Other. I am awed that I can never really attain enough in the classroom to satisfy everyone, not even myself.

But can I really say that as a teacher I hire workers? Except for the admission of overloads, I have no say concerning the occupants of the room. Indeed, I stand in the front of the room, or sit in its rear, and the students enter willy-nilly. The times for class are absolutely prescribed and alternatives to those times are absolutely proscribed. I may not ask students to come early and/or to stay late. And the syllabus, the terms of work, must be adhered to by them. Is the syllabus part of local custom? What part of local custom is included in the syllabus? No, students are not hired; rather, they are for the most part ordered. Universities seek a particular student; they are interpellated, compelled, as it were, by mail. Once they arrive on campus, students' work is mandated—they are commanded. The local custom is here set by the masters. From where are *they* gathered? And what are *their* customs?

Henry Giroux (1992, 242) calls me a cultural worker. Cultural workers are "dedicated to reforming all spheres of public education as part of a wider revitalization of public life." The teacher is no longer isolated in the classroom but acknowledges the place and the power of the classroom in the construction and functioning of the world. To George Counts's question, "Dare the Schools Build a New Social Order?" Giroux responds (1992, 242) with a resounding, "What else?" This transformation will occur as we teachers *cum* cultural workers "resurrect traditions and social memories that provide a new way of reading history and reclaiming power and identity . . . creating new languages and social practices that connect rather than separate education and cultural work from everyday life." I believe education is powerful, but I doubt its power. A character in Chaim Grade's novel, *The Yeshiva* (1976), says, "One can understand the truth in one hour, but to follow it one must struggle a lifetime." When I read Giroux's declamations about the work of teachers, the transformative classroom seems to operate without conflict—indeed, sometimes even without students. Oh, I understand about what Giroux writes, but I cannot without struggle even begin to practice what he preaches. I wonder in what public school classroom Henry Giroux last taught.

In the Fall of 1999, I worked for three months teaching Social Studies—American history—in a private Jewish Day School in the Mid-West. It was the most difficult work I have done in years. For the past decade I had been immersed in curriculum theorizing, and for years I had written voluminously and extensively about the desperate need to rethink educational practices in our schools. In the present instance, I wanted very much to help these Middle School students begin to think like an historian, a term I defined as someone who always knows that there is another story yet to be told. I wanted to use the discourses of popular culture—folk song and original source material—to give expression to the silenced voices in American history and to interrogate some of the stories to which students already gave credence. I wanted learning to be challenging, fun, and, well, empowering. I mean, I wanted to "resurrect traditions and social memories that provide a new way of reading history and reclaiming power and identity . . . creating new languages and social practices that connect rather than separate education and cultural work from everyday life." I mean, I wanted to change the world.

And how should I presume? I miserably failed; I failed miserably. In one class studied Yosef, a highly motivated sensitive young man who actually said, "I enjoy studying with you," and in that same class, studied his sister, Sari, who could joyously miss days of school on end and then not understand why she was unaware what topic we were at present discussing. Yosef finished two books before his sister picked up her first. In one class sat Eli, an intelligent and involved student who suffered from epilepsy and was scheduled at the Mayo clinic for delicate brain surgery to relieve his symptoms; and sat also Rebecca, whose mother has been recently institutionalized and whose younger brother's behavior was so extreme that the family was having difficulty placing him in any school at all. How shall I consider these students' homeworks turned in late and, to my mind, carelessly prepared? In one class sat Jessica Alba, about whom I wrote, "Jessica is highly distractable. The demands of her social life seem to be a priority," and Samantha Rosenzweig, about whom I wrote, "Every teacher deserves a student such as Samantha. Her powerful intellect and imagination stimulate and challenge." How shall I speak to both at once? Or even separately? In one class was Isaac Wills, about whom I wrote, "Isaac has a very difficult time in class. He does not seem to have a sense of inner control and is disruptive and highly distracted;" and Bonnie Wye, about whom I wrote, "Bonnie is one of the most delightful students I have known. Though quiet, she is an enthusiastic learner . . . Her writing displays a sophistication special to middle school experience." Is it American history that would benefit Isaac now or a sense of inner control? Is it a lesson in history or a warm heart that might comfort Rebecca? How do I begin to address the concerns of both children, and those of the fifteen other children in the class and satisfy my own beliefs and practices concerning the American history classroom I imagined? I sit here now recollecting emotion in moments of tranquility and wonder incredulously how I could have ever believed that I might have succeeded!—Or that my success would have been so evident that even I would have been pleased? And if I was pleased, perhaps it was me of whom I was originally thinking: violence, I recall, is to act as if I were alone to act! How could I expect that all of these students should feel the same about school and at the same moment as at that moment of the class in which I was then teaching? And that somehow these students would be intrigued to exactly the same degree and coincident with the present material as I had offered. How could I expect that all of these students would be willing to learn with me, or from me? What was Hecuba to them or they to Hecuba that they would weep for her? What right did I have to demand very much intellectual struggle from them when for many just getting to school was an achievement far in excess of my seventy-mile drive! How could I have expected all of these students to be like me in my idealism and occupy my idealistic image? Transform society? Ha!

Each and every day I was overwhelmed by the insurmountable effects of society upon these children. These were mostly privileged children whose parents could afford tuition to a private school, but their situations were obviously influenced by the personal and family psychological situations and conditions of which I could not even attempt to make sense, much less begin to relieve. Tolstoy had said that all happy families were alike, but that every unhappy family is unique. He was wrong: every happiness, too, is different. Weren't Yosef and his sister both seemingly happy and yet the product of the same home? Or must I abandon all belief in happiness and assume it to be yet another mask. There was something intrinsically wrong with the whole notion of the classroom as it architecturally, structurally, socially, intellectually, and academically existed. How could I possibly deal with all of these personalities and still engage in American history learning? There were days when all I wanted to do was to teach these children some sense of inner control. There were days when all I wanted was to get out of there alive. Oh, I wanted them to learn some American history; I wanted their stories to assume rightful places and add to the discourses of American history; I wanted their stories to intervene in the monolithic American legend. I wanted to be a great teacher. How could I presume? How many eyes must one man have before he can see people cry? I have rarely felt as isolated in a classroom as I did during those months. What influence I might have had very little to do with what I taught. Local custom? I knew none.

Cultural workers must discover the local custom because for the most part, as James McMurtry sings, "I'm not from here, I just live here." Giroux valorizes this aspect of the pedagogical encounter as border crossings. And how should I presume? And I wonder, what would serve as my passport? Aren't border crossings also interpretable as trespassing and border crossers as illegal immigrants? Woody Guthrie laments that those who cross may be known by no names, except "deportees."

I sense that I have begun to complain. I do not like to hear myself complaining, but I tend to complain when I am insecure. Talking, I know, of which complaint is a species, is far easier than doing. Shammai said, "Set a fixed time for your study of Torah, say little and do much, and greet all men with a pleasant countenance." I said too much in that classroom of middle schoolers. I am certain that I did not always meet them with a pleasant countenance. Complaining, like Winnicott's fantasy (1971, 26), is saying much and doing little and represents a waste of time and energy better put to use finally, well, within the classroom. And when I complain, I simply invite Job's comforters to relieve my complaint. It is, finally, never to them that I must appeal: "Is my complaint directed to a man?" Job asks. His comforters, as mine, must always fail for they will simply assume a simple cause of complaint and offer an absurd and simple relief for it.

Actually, when I complain it is not to them I wish to listen. Rather, I complain to have them listen to me: why else complain? I would really stop complaining, but first I would recognize of what my complaints are a symptom. I must resolve—or begin to begin the resolution of the conflict between what I say in the classroom and what I do in it—to begin to begin to resolve the tension that exists between my devotion to my own study and my advocacy of study in the classroom. Rabbi Ishmael said, "One who learns in order to teach is afforded the opportunity to learn and to teach. Whereas one who learns in order to practice is afforded the opportunity to learn, to teach, to observe, and to practice." I teach in a school that educates teachers: students demand that I teach them "how to teach." "Tell me what to do," they plead, "and we will do it. Just please, give us direction." I now wonder, for what am I educating them to do: to teach or to practice? If the former, I teach a craft but if the latter, I offer a vision, an art, a way of life. Rabbi Ishmael's statement suggests a qualitative difference between learning to teach and learning to practice: I think what I have always sought in the classroom is ethics and not pedagogy.

✡

Hillel

"IF I AM NOT FOR MYSELF, WHO IS FOR ME? IF I AM ONLY FOR MYSELF, WHAT AM I?"

The complications that arise from the complex relationship between myself and students in the classroom are not uninteresting—or uneventful. I do not refer here to how I personally feel about the particular students or they about me. Of course issues of, perhaps, personality though—these sensibilities must be taken into account in any deliberation about the classroom. Nor do I refer here to matters concerning the boundaries erected in the classroom based in hierarchical orders previously established in and by the world—questions of power relations. Though of course, these borders must be taken into account. Nor are the complications of which I speak questions of subject matter—traditional concerns of curriculum. Though of course, these issues too, must be taken into account. I refer, rather, to the essence—the very purpose or necessity for the classroom and our places in it: what we to do in there, alone and together? Hillel has said, "If I am not for myself, who is for me? If I am only for myself, what am I?" Maimonides stated that Hillel refers here to the necessity of "myself" to assume its freedom and to bestir the soul to virtue. Maimonides asserted that because there is no external cause to which our actions might be attributed, then all choice must be attributed to personal freedom. Only *I* can choose for me. In contradistinction to significant aspects of American Puritanism, Maimonides proves that there is no predestined order mandating our actions: we are absolutely free to act and we must assume responsibility for our deeds. Maimonides's interpretation of Hillel's famous statement makes me absolutely responsible for my prerogatives in the classroom—these are my choices. It is myself, I, who must take responsibility for this classroom— the local custom is first mine, thank you very much. And for Maimonides, once the individual has chosen specific deeds, then that individual must self-consciously acknowledge what s/he has now become: having now made my choices, this is the teacher I am.

I think that for Maimonides the interpretive emphasis is the pronominal "*I.*" In this regard Maimonides seems inclined to understand Hillel's dictum as pertaining to the relationship between individual freedom and the development of character. How have my choices made me what I am! Maimonides talks not unlike some therapists I have frequented who kept asking me how I felt about the decisions I had made and why I had made such decisions. I have spent many hours in such discussions and I am not unsympathetic to Maimonides's interpretation.

But I would now prefer to understand Hillel as giving voice to the constant tension between the needs of the self and the demands of the larger community, in my case, the space that separates my *beit mikdash* (house of study) down here in my basement and my university classroom up there on the hill, between my desire to learn and my desire to teach, the former undertaken in quiet solitude and the latter amidst noise and tumult. The former I accomplish for myself; the latter perhaps, is what I am, or how I appear in the world. The former derives from the desire to invent a story that becomes the world of my dreams and the latter exists as the story I tell about the world that I have dreamt. I live amidst the daily tensions that arise from the contradictions deriving from the conflicting demands attendant upon learning and teaching. Do I teach what I have already learned, or do I teach so that I may learn; do I learn while I teach? Do I teach while I learn? It seems to me that the responses to these questions determine the type of lesson plan I prepare and determine the type of teacher I will be. If I already know what others must learn, then don't I acknowledge that knowledge is quantifiable? If I already know what others must learn, then aren't all of my questions specious—I know what the answer is and it is merely my chore to lead students to that answer.

✡

In the dark, stuffiness of the cheder, the Rabbi stands before the class of children. The youthful charges, some not more than five or six, sit on the benches with their feet dangling several inches from the floor. The Rabbi is tall and his beard is wild and flowing. He wears the traditional garb: black trousers and what seems to be a permanently soiled white shirt, black coat which flies up behind him as he paces back and forth before the students; on his head sits a black *kippah* (skullcap). The young students sit silently, expectantly and terrified. He stares at his charges and in a stern voice asks, "Class, can any one tell me why the first book of the Bible, *Bereshit*, [רישאב] begins with the letter *gimel* [ג]?" Confused and panicked, not a child moves, nor offers a response. Each sits with his head down trying desperately to avoid notice. "Well," the Rabbi thundered, "Can anyone tell me why the first book of the Bible, *Bereshit*, begins with the letter *gimel?*" He stares with wild, almost crazed eyes at the frightened children sitting quaking in their seats. The silence roars. "Can no one give me an answer?" the Rabbi roars. Suddenly, from the back of the room a timid hand goes up. "Yes," the Rabbi says, "can you please tell me why the first book of the Bible, *Bereshit*, begins with the letter *gimel?*" In a hushed and tentative voice, the child responds, "But Rabbi, the first book of the Bible doesn't begin with the letter *gimel* [ג]; it begins with the letter *bet* [ב]." The other children sit horrified and await tremblingly for the wrath of the Rabbi to come down on their poor colleague. The Rabbi stares out at the brave child for another ten seconds, and he says with a soft lilt in his voice, "Well, that's one answer!"

I have always loved that story. In that bleak and dark classroom, conversation would never end. Yes, I could respond, that is one answer. For every answer, another question. There is, of course, a danger for me in that open classroom: If I teach students how to ask questions, then how can I stop them from asking me questions for which I have no answers. How shall I control the classroom when the world might be in there? And if answers are not readily forthcoming, can I consider myself yet a teacher? What should occur in the classroom that learning would take place? What is the learning that must take place in the classroom? Alternatively, what is the motive for a type of learning based in objectives? But no, that is a false question, because I already know that answer.

Present issues of curriculum swirl about issues of control: control of children, control of knowledge, control of the dreams. Control—that is the only answer in the classrooms of the United States. It is comforting to know exactly what material must be covered; it is satisfying to know the material has been covered. But I wonder, if there were no objectives, then what would we do here daily? How will we know when the class or the day or the year should end? I do not think that these are matters to be dealt with by epistemologists nor efficiency experts nor even curriculum developers: I think that these are issues of ethics. These are issues that help me answer Hillel's third and equally difficult question. "And if not now, (and especially during this Schmettah)[2] when?"

✡

"IF NOT NOW, WHEN?"—A TALMUDIC DIGRESSION

The doctors of the Talmud were forever concerned with the tensions that develop between the life of scholarship and the daily life in the physical world. If Torah study was the most important activity, then everyone should be engaged in it, but then, how would the world keep on keeping on? The relationship that is maintained in society between the scholar who learns for the sake of learning and remains free from the requirement of production and the world, which supports the scholar's learning with its productivity, is a matter of serious concern to the Rabbis. Long before H. G. Wells wrote *The Time Machine*, the Rabbis must have been troubled with the development of a tiered society where the intellectual and cultural elite live off of the labor of the workers. What is, the Rabbis queried, what should be, they asked, the relationship that exists between the scholar in her study and the society in which the scholar works? It is apparent that society offers the scholar the materials of daily subsistence, but what does the scholar offer the society?

This tension plays itself out in many aspects of American society, but at the university it is most obviously found in the discussions amidst faculty and administration and Boards of Regents concerning the conflicting obligations of teaching and research. The tension exists as well in the perennial conflict between university and public secondary school faculties where the latter are often perceived (and perhaps, often perceive themselves) as second-class educational citizens because their primary obligation is to teach. The same antagonism exists, I think, between secondary and elementary school staff because the former situate their work in the rigor of the academic disciplines and the latter, well, they care for children. This strain is exacerbated by relations of gender in the elementary grades.

This tension between the practical and the scholarly modes of the teaching profession also inhabits the tone in the ubiquitous discussions concerning standards and national curricula. What should a teacher do and teach? How should a classroom be constructed to prepare such teachers? Are teachers taught to teach or to practice? What is meant by practice?

How do we valorize the construction of knowledge—even transformative knowledge—unless we are actively engaged in it ourselves. But what could be the relationship between the scholarship I practice down here in the basement, the scholarship I espouse at the university, and the daily activity in the classroom? If my learning is always first for me, if when I study it is me who is fulfilled, then how can I stand face to face with students? What is the connection between my study and the world, which is too busy to study.

I read about Rav Rehumi:

> Rav Rehumi could always be found before Rava [a teacher] in Mahorzah, and used to go home every Erev Yom Kippur. One day the [portion of Torah under study] allured him. His wife was waiting for him:
> "He's coming now! He's coming now!" He didn't come.
> She lost her senses, a tear fell from her eye.
> He was sitting on a roof—the roof opened underneath him and he died.

Rehumi is the quintessential scholar. All year he studies Torah; it is his life and his delight. Rehumi does nothing all year but engage in study. And while he is engaged in the learning of Torah in the *beit mikdash* (House of Study), the world is maintained by those who are not engaged in such study. After all, Rehumi and everyone else must eat. And so it is Rehumi's wife (who else in this patriarchal world?) remains at home and carries on the business of daily life, in effect, maintaining Rehumi by ministering to the world and to his world that he may remain in study.[3] His study is wholly for himself; it has no end because he does not leave it. He does have to leave it because it has no end. But . . .

On the eve of every Yom Kippur, Rehumi returns home to fulfill familial obligations. Since it is true that on Yom Kippur atonement may be made for sins against one's fellow human beings only by personal engagement, Rehumi can only fulfill this obligation by returning home. And so every year on the eve of Yom Kippur, Rehumi returns home to ask forgiveness of those who must continue to suffer that he might continue to study. Rehumi must acknowledge that his study is maintained by the hardships and sufferings of others. This particular year, however, he must have been especially entranced in his Torah study and he neglected to return home to his wife. Rehumi had failed to meet his human obligations and to acknowledge his wife's suffering and his own guilt. Her tears fell and Rehumi died. Her tears destroyed him. She cried and he died.

There is a tension here—study is valorized as all-important, but not to the total exclusion of the world. Rehumi's single failure destroys him. Rehumi studies all year; his learning is so important (to whom? for what?) that others are willing to work to sustain him in the *beit mikdash*. Rehumi may study only if someone else agrees to engage in the daily activities required to maintain not only their own existence but the existence of the scholar. Rehumi may study because others not only do not study, but must, rather, labor and suffer that Rehumi may continue to study. We know Rehumi studies out of love; he parts from it only by compulsion. Who, or what, does Rehumi love?

To the question: How could study be so valorized even in the absence of deeds? The Rabbis offer an answer. The scholar and the world exist in a precarious equilibrium and the balance is constantly in danger of being upset. The Other's suffering endangers the security of the scholar's world—endangers the security of my world. This story blames Rehumi not for his almost constant study, but for his single failure to acknowledge the world by not returning to it even on this Day of Atonement. Even study is not an excuse to avoid the encounter face to face.

It is never easy to come out of my basement office and walk into the classroom. Immersed in my books I am awash in a holiness. Immersed in my books I am safe from the world. I have written:

> I stand before a class and I am unnerved by their beating hearts. I know that the students look to me as the one who knows . . . They are terrified of my power. I am terrified by my power. I am frightened that all of my knowledge is inadequate to the task that abides in the classroom. I am terrified that I have nothing to teach these students. I am terrified that I will not know what to teach these students. I am terrified that I am alone in this classroom. Truly, I wonder, what exactly is my task as I stand in front

of this classroom and how will I judge the accomplishments of the students and of myself from the position at the front?

The insecurities that arise from the incompleteness of my understandings are hidden down here in the basement; down here in the basement the incompleteness of my understanding is the motive to stay down here in the basement. When I go up, what shall I say? When I go up, what shall I be? I can remain down here and avoid the necessity for all actions for as long as I remain down here. But who will then assume my burdens up there?

But, I protest, I am not just afraid down here. I am bravely dreaming a world!

I think this story of Rehumi epitomizes the conflict I have lived with in the teaching profession for most of my life in yet another way. If learning is undertaken for its own sake, then what are the needs for academic standards? Standards assume an end to study; a measurable and acceptable level of competence that enables some graduation to another level. But if study for itself is the end, then there is no subsequent next step necessary and therefore, no measurement of success required in order to advance. Indeed, there is no advance. There is here a motionlessness to the activity of study. Or perhaps, there is a leisure to it that obviates the need for standards and grades and all of the other hierarchical architectures that I despise of the school. What the students do not know today, by their questions they may know tomorrow; study is ceaseless. As learning for its own sake is an absolute activity, standards become irrelevant because there is no finite end to which study must lead. Study is itself an end. If study is its own end, then how might a lesson plan be ever structured to fulfill objectives? What objectives? "Well," the Rabbi avers, "that's one answer." In this response, the Rabbi offers neither direction nor end to study; lesson plans are inevitably built upon the compulsion of both.

I have written:

> My eight-year-old daughter, Emma, sits peacefully on my lap. She is reading Perrault's version of Cinderella. It is beautifully illustrated. "Emma," I ask her not innocently at the tale's end and because I am a teacher, "What do you think that story is about? I mean, why do you think the author wrote that story?" And my daughter looks up at me earnestly and says, "Well, Dad, I think the story is a metaphor for the evils of a burgeoning capitalism which encourages a certain acquisitiveness in the bourgeoisie who, aspiring to wealth, are willing to abandon and even exploit their natural allies, the workers. And furthermore, Dad, *Cinderella* doesn't say very nice things about women and perhaps, you shouldn't share it with other children, especially my little sister."

If the text is inexhaustibly interpretable, then there is no definitive meaning that must be learned. How should I then assess my daughter's reading skills based on her response? What shall I say of her learning? Of her comprehension? Moses, to whom the law was given at Sinai, asks God why God takes the time to put all of those tittles and crowns atop so many of the letters of Torah. And God responds that there will be someone who can brilliantly interpret each of those seeming ornaments. "Show me that man," asks Moses in awe and amazement. God commands Moses to turn around. Moses does so and finds himself in the rear of the academy of the great sage, Rabbi Akiva. Sitting in the last row of benches, Moses hearkens attentively to the conversation, but understands not a word. Then a student questions Akiva concerning a particular interpretation of Scripture, "How do we know this?" the student asks and Akiva responds, "It is the law of Moses." Hearing this, Moses is relieved. Interpretation of text is illimitable; it is not the text but the interpretation of it that remains paramount. All meaning resides here.

Text is the motive for interpretation and interpretation is endless. Which is a relief to me who has been engaged all of my life in textual study. Most school texts, I note, however, are written to require absolutely no interpretation. That is their argument: knowledge is presented in them as seamless and complete to be simply learned rather than cogitated over. How might I engage in the processes of study when only its products are valorized? How do I teach the beauty of study when the indisputability of the answer occupies the classroom as scheming as a poltergeist? How do I teach interpretation when so many forces of certainty are arrayed against me? But if I choose not to teach interpretation, then what kind of teacher am I? How would I then respond to the student who says, "Be a teacher. Teach me." "Well," the Rabbi avers, "That's one answer."

IT IS NOT IN HEAVEN

The Torah (as are all texts) is meant for study. But the Torah, according to the Talmudic doctors, must be first interpreted—and this, Talmud suggests, must be undertaken in accordance with local custom. Or rather, though there are hermeneutical principles that valorize technique, interpretation may not disregard local custom; indeed, as we have begun to see above, all is practiced according to local custom. That is, though I may maintain my own opinion, I must recognize it as only my opinion and not law. There is a story: Rabbi Eliezer disagreed with the rulings of his fellow Rabbis. Despite Eliezer's insightful arguments, the Rabbis did not follow his opinion.

[Eliezer] said to them: "If the halakhah is in accordance with me, let this carob tree prove it." The carob tree was uprooted from its place one hundred cubits—and some say four hundred cubits. They said to him: "One does not bring proof from a carob tree." He then said to them: "If the halakhah is in accordance with me, let the channel of water prove it." The channel of water turned backward. They said to him: "One does not bring proof from a channel of water." He then said to them: "If the halakhah is in accordance with me, let the walls of the House of Study prove it." The walls of the House of Study leaned to fall. Rabbi Yehoshua rebuked them, [and] said to them [the walls]: "If Talmudic Sages argue with one another about the Halakhah, what affair is it of yours?" They did not fall, out of respect for Rabbi Yehoushua; but they did not straighten, out of respect for Rabbi Elazer. He [Elazer] then said to them: "If the Halakhah is in accordance with me, let it be proved from Heaven." A [heavenly] voice went forth and said: "Why are you [disputing] with Rabbi Elazer, for the Halakhah is in accordance with him everywhere?" Rabbi Yehoshua rose to his feet and said: "It is not in heaven."

What does "it is not in heaven" [mean]?

Rabbi Yirmeyah said: That the Torah was already given on Mount Sinai, [and] we do not pay attention to a [heavenly] voice, for You already wrote in the Torah at Mount Sinai: "After the majority to incline."

Despite the brilliance of Eliezer's reasoning, despite even the support of the heavenly voice, the Rabbis reject Eliezer's arguments. "It is not in heaven" means that though the words were given on Sinai, their meaning is an earthly matter. It is not that the words are unimportant; rather, the words are crucial; it is just that they are opaque. It is the meaning to be fashioned of those words to which attendance is paid and that meaning is local.

That is a comfort to me as I sit with my own and my daughters' meanings. However, when I go to where the reference is made to Torah (the words)—when I read in Deuteronomy 30:12, I am surprised by what I discover:

For this commandment that I command you today—it is not hidden from you and it is not distant. *It is not in heaven* . . . Rather, the matter is very near to you—in your mouth and your heart—to perform it [italics added].

The reference in Deuteronomy suggests that the mitzvot (the commandments) are not strange to human behavior, but are rather intrinsic to an understanding of what it means to be human. The meanings and matter of the commandments are not strange, nor distant, nor ideal. Torah states: The commandments are not in heaven—they are here, on Earth to be acted upon in daily life in my daily communings with the world, which includes other beings. The meanings of the mitzvot are not obscure but apparent. According to Torah, it would seem that interpretation must be relatively unnecessary. There is nothing local about it at all; indeed, meaning is universal.

I am surprised, then, by the Rabbis' mild misreading of "it is not in heaven." In their view, the phrase valorizes interpretation as a human and local prerogative. The meaning of mitzvot must be decided here and not in heaven. For the Rabbis, "it is not in heaven" makes interpretation the central activity of study, though the end of study is clearly performance. Torah has said that the mitzvot are not about heavenly authority, nor even scholarly activity, but about earthly responsibility. A relationship to God is a relationship to human beings. Torah states that these mitzvot are not obscured, as is the top of Sinai by clouds of smoke: they are clear and readily understood down here on the earth. The mitzvot, however, are to be performed. They must be enacted—I am Holy and you should be Holy—but the meaning of the mitzvot issues not from heavenly edict but from earthly exegesis. The meaning of the mitzvot derives from human action—interpretation—and must be negotiated in personal encounters—Talmudic-like discourse. "It is not in Heaven" means we are ultimately free. The Rabbis mild misreading offers me the opportunity to link interpretation and action and to make interpretation an ethical presence. I wonder at how our classrooms might alter if the teacher would assume such a stance. "What does this mean, teacher?" "It is not in heaven . . . but in your mouth and your heart to perform it."

But when I seek the second reference, "After the majority to incline," the Rabbis urging to subscribe to majority opinion, I discover a strong misreading. Exodus 23:2 actually says, "Do not be a follower of the majority for evil; and do not respond to a grievance by yielding to the majority to pervert [the law]."

In their response to Eliezer, the Rabbis have turned the Biblical negative into a rabbinic positive: Exodus says *don't* follow the majority into evil, but the Rabbis contend that, in fact, one *must* follow the majority in interpretation. This strong misreading is effected to reinforce the notion that all is according to local custom—it (meaning) is not in Heaven. Interpretation defines the mitzvot; human beings perform interpretation and meaning is made possible through the awareness of local custom. Of course, the Rabbis have valorized the scholar's interpretation; they are scholars and they have invested themselves with ultimate authority! But they inadvertently valorized also the interpretation of the many scholars who would succeed them—some of whom reside in our classrooms. Even the Heavenly voice might not cancel these interpretations. I welcome the dissent. But in our schools the question is still asked our daughters: "Which of the following most matches the author's purpose in writing this passage?" I urge Emma and Anna Rose not to fill in any bubble, but instead to pencil in the reply: "It is not in heaven! After the majority to incline." And what machine will read their response?

I think the Rabbis meant to ensure that interpretation must have human application and that interpretation must address local custom. Authorized meanings must be livable, which is to say, ethical.

✡

I think it specious to say that we work for our students. Over the last ten years—the years, of course, that I have been at Stout and from which I am now celebrating a sabbatical—there has been a great deal of this type of discourse at the University. Students are thought of as "our customers" or "our clients;" we provide them, we are informed, with a service and we are often urged to act accordingly—to give them what they want; treat them with greater deference; try not to antagonize them and to make them happy. Remember, we are cautioned, that we are here to serve them. Furthermore, there has recently been a resurgence of talk that curriculum must serve the national interests; as a teacher I am enlisted as a soldier in the cause. I train workers. Now, as my customers, my charges must be happy, but as future workers they must be ready. I am the educational drill sergeant. They must meet muster.

I have never understood this type of talk. Perhaps I was always plagued by the identity of the student I was meant to serve: there were, after all, so many desires out there in the classrooms. Perhaps I did not know what product I was supposed to offer for some kind of purchase, or what I was to put on display for inspection and acquisition. What part of my knowledge could I possibly make available for hire or purchase within the scant parameters of time and space set by the schedule and the environment? Who, indeed, had set the nature of employment by which I was to operate? What local custom was operating here? If these students before me were customers, then were discomfort and irritation banished from the room? I do not think that learning can take place without it. What could I possess that everyone in the room could concurrently desire? Rabbi Baer of Radoshitz once said to his teacher, the Rabbi of Lublin: "Show me one general way to the service of God." The tzaddik replied: "It is impossible to tell men what way they should take. For one way to serve God is through teachings, another through prayer, another through fasting, and still another through eating. Everyone should carefully observe what way his heart draws him to, and then choose this way with all his strength" (Buber, 1947). This would be a classroom neither of purchase nor sale but of individual pursuit. Where would I stand in such a classroom?

Perhaps it was I who was meant to be up for sale? I wondered who I was supposed to be and how had I been represented? How would I know when I had been purchased? And do I come with a warranty? The questions are absurd, of course. If I aim to please, I must miss my mark for we cannot know what we do not know—how can a student name precisely what they desire from each class? How can students know in advance what will give them satisfaction? We must move about in the classroom in confusion and direct ourselves always toward the ever-receding light. Rabbi Moshe interpreted the verse in Exodus 27:20, "Now you, command the Children of Israel, that they may fetch you oil of olives, clear, beaten, for the light . . ." to mean that "we are to be beaten and bruised, but in order to glow with light" (Buber, 1947). In a world of endless interpretation and limitless knowledge, not the achievement of satisfaction but of dissatisfaction ought be our purpose.

I prefer to think that the classroom is a place in which we might collectively pursue questions in an atmosphere of relative security. In the classroom we might negotiate our contested meanings, but nothing need be purchased, nor need anyone be responsible for marketing anything. I mean, what does learning have to do with selling? I cannot name what it was I got from my teachers, though I regularly received graded report cards. How can I even immediately know what I ever got from my teachers? Perhaps the classroom is rather not where one gets or even where one purchases, but rather, the classroom is where one takes. It is our activity and responsibility that acquires. Rabbi Shelomo (Buber, 1947) said to his disciples: "After death, when a man reaches the world of truth, they ask him: 'Who was your teacher?' And when he has told them the name of his teacher, they ask: 'What did you learn from him?' One of the disciples cried: 'I have already prepared what I shall say in your name. It is "May God give us a pure heart and pure thinking, and from our thinking, may purity spread throughout all our being, so that in us the word may be fulfilled."'" "Before they call, I will answer." But my classrooms are so quiet. I imagine the students await the sales pitch.

And what should be the established standard, I wonder, if, as the student above said, it is purity and good deeds that one learns in the classroom? How, indeed, might I teach this and how might that be measured? I am aware that the ideal classroom I have constructed never truly existed.

It was all a dream. Woe is me! Alas, for us! Rather, the classroom I now would like to occupy is a response to the classrooms in which I have lived. I have come to appreciate the extremely difficult social, cultural, racial, and sexual maelstrom that whirls angrily, and often violently, in the classrooms of the United States and in which we dramatically and almost tragically twirl about with terrifying and bewildering force, willy-nilly. This seething vortex subverts the freedom that the classroom might champion if it could only valorize the pursuit of the question rather than the accomplishment of the answer. I think that it is the question that will save us and not the answer. The ability to ask questions and to pursue answers, but not to achieve them, seems to me to be the essence and center of the classroom and I begin to suspect that it is Rabbinic Judaism that institutionalized the question at the pedagogical center of practice, even at the center of canonized texts. But the students sit in fear and in malaise desiring some purchase to validate their presence. Or to certify their Being. They are silent and surly. To their question "why am I here?" I have no answer.

But down here in the basement on this sabbatical I am away from these fields. It is relatively quiet, except, of course, when the children are home.

The Talmudic Rabbis themselves authorized the freedom that occurs with the establishment of the question at the center of textuality. That opening was redemptive and pedagogical. Following the destruction of the Second Temple in 70 C.E., when without the unifying force of the central locus of worship and the essential practice of the sacrifice, the future of Judaism looked bleak. In their effort to reinvent Judaism, the Rabbis assumed authority in a great many areas of Jewish practice once held by the now irrelevant priests. The Rabbis situated leadership roles in themselves and canonized certain texts; they decreed that the superfluity of the priests might be obviated by transferring their functions to the people. They read into those now canonized texts traditions whose practices would make possible the development of the Jewish people into a nation of priests. Study, reading God's will into history, and prayer, a commitment to humility, replaces the traditions of the sacrifice that were no longer possible with the loss of the Temple. Thus, the commandment that "you should tell your children" about the deliverance from Egypt grew into the exercise of the Passover Seder and led to the institution of the question as the impetus for the telling of that story. In this authoritative, brilliant innovation, the Rabbis also included a corrective to their own hegemonic act; the Rabbis inserted into the traditions of Judaism the subversion of their own authority by demanding that questioning be integral to the process of study. The practice of study originates in the utterance of the question. Ouaknin (1995, 87) says that "Talmudic thought is the thinking of the question, and it is not mere chance that the very first word of the Talmud is a question: *Meematai* (From what time?)." For central to Rabbinic Judaism was tradition interrogated. Indeed, the loss of the Temple demanded the inquiry of texts for the purpose of discovering new tradition. Hence, no question is answered in the manner in which it is asked and all answers are meant to provoke further questions. "The answer," Blanchot states, "is fatal for the question." My students, I remember, have been taught to demand answers.

The truth would never set us free, but perhaps, the pursuit of truth might approximate the exercise of a freedom great philosophers have only described. It was, of course, John Dewey who argued that the relationship between democracy and education was symbiotic; without education, democracy was impossible, but without democracy, education was reduced to indoctrination and social control. Such education would be conducive to totalitarian states—the Hitler youth a prime example—but not to democratic societies. The pursuit of truth demanded questions and not answers. And so the reduction of students to the status of client for whom answers must be provided on demand and the designation of teachers as salespersons who must serve at the caprice of the student *cum* client reduces our role to that of a barker at a carnival. I set up the booth and display conspicuously the prizes that may be attained; I blow up the colorful balloons and tack them to the board. Then I stand at the counter and I holler coercions at the passers-by: "Hey, try your luck. Everyone a winner. It's easier than it looks."

At these carnivals, attendees walk and are hailed (I like Althusser's notion of *interpellation*—we were defined by that hailing) by the rubes standing at their game booths: "Hey pretty girl, try it once for free and win a stuffed doll;" "C'mon, missy, give it a try, you can do it! Free prize no matter what your score." The first game is, as they say, free, and always gathers in a free prize. And then the dollars start rolling between my pocket and my daughter's desire and the rube's avaricious hands. Emma and Anna Rose would play forever: is it the game or the prize that entices? Sometimes I consider that it is the latter, which attracts. After all, their rooms are stuffed with these creatures; there seems to be no end to this desire—I wonder if they are lonely? But maybe it is as well the challenge of the game; and at the outset, the inability to fail. Everyone a winner! Could it be I am finally mistaken—officiously self-serving—to distinguish myself from these carnival rubes? Perhaps I do need to offer prizes. "Hey, little girl, try this book. First few pages are free. Everybody's a winner. Take home a stuffed animal!"

But I always wonder what it is I am supposed to be selling. What do I have that anyone might desire? I suppose that students must always answer that for themselves: I am not at all certain what I have to offer. Rabbi Mendel said, "I became a hasid because in the town where I lived there was an old man, who told stories about tzaddikim. He told what he knew, and I heard what I needed" (Buber 1947). I must first and always think of myself as a learner and learning has traditionally been, if anything, about acquisition. I stand in the classroom acquiring! I take what I need. I tell what I know and they take what they hear. But though constructivism might be prevalent in the academic community, it appears absent from my classroom. Students sit in the classroom passively. Students ask me, "What did I get?" as if learning was conferral. I am reminded of Arlo Guthrie who, sitting on the "Group W" bench when asked "What'd you get?" responds, "I didn't get nothing. I had to pay fifty dollars and pick up the garbage." To the students' query, "What did I get?" I might respond, "I honestly don't know. What *did* you get?" Would they look at me with disbelief and suspicion? Where did they learn this subservience? But I know they have learned it in the classrooms they have previously occupied.

My friend Peter Appelbaum suggests that perhaps the activity of the classroom might be likened to poaching—a clever taking of what has been selfishly fenced and guarded. The more ingenious the poach, the more joy in the activity. Roald Dahl's novel, *Danny, the Champion of the World*, portrays a world in which poaching is a benign activity engaged in by everyone during the pursuit of their daily lives. They only poach what has been selfishly withheld and they harm no one in the endeavor. Acquisition is entertainment. Seen in this light, learning might be conceptualized as an adventure marked by acumen. What the student takes from the teacher does not diminish the teacher, but fulfills the student. "My field is filled with game; it is mine, but you may take what you will. It will change you, I promise, and not diminish me." The teacher must make certain, however, that the fields are filled with objects to be desired. "Hey, pretty girl, take what you will. Everything's for free. But you gotta take it yourself." Lawrence Kushner suggests that God intended Adam and Eve to eat of the fruit of the Tree of Knowledge so that they might have to leave the garden and embark on life. Take, but do not ask my permission. It will change you forever. We become in the act of taking.

As I learn, I assume different stances in and towards the objects of the world—learning is thus as much about issues of ethics as about issues of academics. Ethics is the stand I take in the world before its myriad objects; ethics is how I confront the Other and is at each moment an evocation of my character. The idea of being a learner makes possible certain relationships in the world and when I stand before students—when I stand with students—I am in relationship. Since I do nothing but stand with students, then I suppose that what I teach must be ethics. Teachers must be wisdom seekers—we are, after all, educators, and wisdom is, in large part, not what I know but what I know I can know in pursuit. It is the students before me that make me the teacher. I must remember that as I stand here lecturing. Wisdom is not merely the recognition that I do not know everything, but the acknowledgment that I seek knowledge—hence, the Rabbinic institution of the question. I offer not knowledge but wisdom—it ought to begin with a great silence. What would it sound like?

A Talmudic scholar, Davis Weiss Halivni (1997), offers me an interesting way to think about this concept of silence. Perhaps, he suggests to me, silence resides in the space between the wisdom that I seek and the knowledge I already possess. The gap between the two remains interminably wide and inspires my humility. Halivni argues that the Torah given to Moses and the Hebrew people at Sinai upon the Exodus from Egypt had over the years fallen into disregard and disuse. After the destruction of the First Temple in 586 B.C.E., Ezra the Scribe became the redactor and teacher of that maculate text. Ezra rescued the Torah from the oblivion into which it had fallen as a result of the people's neglect of it. Peter Ochs summarizes Halivni's argument:

> The Talmudic tradition is . . . that the children of Israel sinned which means that they did not always maintain the temple or the Torah scrolls it contained or the teachings within them. Halivni reasons that if Israel did not properly maintain the Torah, then there is no reason to assume that the Torah text and traditions were correctly transmitted through the period of the Kings to Ezra. The implication of this assumption is that Ezra receive what [can be called] . . . the "maculate Torah": a not-immaculate statement of God's words.[Halivni 1997:xiii]

It was Ezra's task, then, to correct this maculate text and to leave behind oral instructions on how to read the text to emend this corrupted yet sacred text. These instructions are transmitted through generations and by education. Thus, according to *Pirke Avot*, 1:1, "Moses received the Torah at Sinai and transmitted it to Joshua and Joshua to the Elders and the Elders to the Prophets and the Prophets transmitted it to the Men of the Great Assembly." Hence have the Rabbis established the authority of a group of teachers as the transmitters of the corrupted text. And just as significantly, they present to us a method, a pedagogical technique:

interpretation of the text. What remains in silence, however, is the text's meaning. As it is maculate, the text's meaning is always in the future. The classroom begins, perhaps, in this paradox. The Men of the Great Assembly—no longer prophets but lesser—must pass on the tools for the continual emendation of the maculate text. Since they do not know what that meaning may be, they must ask questions of the text in the context of the world in which they live. *All is according to local custom.* The silence of which I spoke above is provoked, perhaps, by the humility that derives from the awareness that the task in which we engage is illimitable and its end indeterminable. Nevertheless, our quest affirms life. I am holy and you should be holy. The text is maculate and I am only human. The end of interpretation is only a beginning.

Is that why study is so holy?

Perhaps here is one of the causes of the conflict between the scholar and the population. For those who do not work at emending the text, or do not know that it requires emendation, remain bound to simply comply with it—or be made to adhere to it—and must do so at the direction of those who assume responsibility for the text's emendation. I think there must always be strong resistance to this yoke. All people may be intellectuals, says Gramsci (1971:9), but only some assume the social role of public intellectual.

> All men are intellectuals, one could therefore say, but not all men have in society the function of intellectuals . . . [whose] activity is weightier towards intellectual elaboration [than] towards muscular-nervous effort.

Schools tend to teach from immaculate texts. Students are given answers and denied questions. They are denied intellectual function. They inevitably resist.

Nor in this scheme may teachers be considered social intellectuals because the texts they employ are considered immaculate. Isn't that why they are reproduced *ad nauseam* in the form of worksheets, homework assignments, and standardized texts. The interpretative effort of the public intellectual—the elaboration of textual emendation—is criticized as mere mental gymnastics, elegant and graceful but little more than show—or ridiculed as masturbatory, self-absorbed activity satisfying individual desire but judged ultimately sterile, enjoyable but not fertile. The work of the teacher, the public avers, should be practical and utilitarian. There must be a discernible and viable product. Teachers ought to be society's servants and not its intellectual leaders.

I am a public intellectual whose activity is weighted towards intellectual elaboration. Mine is a social role. But in the quiet of the basement, I become aware that perhaps what I have always taught are only the tentative responses of people who before me had earlier sought answers to queries similar to our own about the world. They offered not *the* answers but *their* answers. The questions remain. Even as I authoritatively lecture, I yet stand in the midst of questions rather than answers. It is in pursuit of our own questions and not the answers of others by which we are compelled to continue. We must always deal with the maculate text and we can never know if and when we achieve its immaculateness. Finally, I suppose that what we *know* is never about the world, but about our relationship to the world and that is always changing and never wholly within our control. That we continue to learn is an acknowledgment of the incompleteness of our understanding, of the maculateness of the text. I pursue understanding about those things in which I have interest; those things are always present to me in the form of questions, and my learning constructs answers to those questions even as it produces new questions that require further study. There is no light at the end of the tunnel because there is no tunnel. There is only the world. But when I take this stance before my students, they wonder where I am.

✡

When people walk into the classroom—especially as mature as they are by the time they get to me—they are products of local customs with which I am thoroughly unfamiliar. I work in an environment replete with local custom with which I have had little to do in its formation. These students are all gathered from different places. I suppose custom changes like the earth moves—in micrometers and over long periods of time. As a result of my work, I might even become a part of the change of *minhag* (the Hebrew term for custom), but of course, even as I don't feel the earth move (except in the movement of earthquakes that results too often in tragedy), I don't feel the change in *minhag*.[4] I just keep on keeping on.

Perhaps my sabbatical leave is an attempt to stop that ceaseless but almost imperceptible movement so that I might know where I stand. But when I exit my cave, I must take care to neither dismiss the world's efforts, nor look with contempt on its reality. I must know and observe local custom and understand my place in it. From within local custom, wisdom might be discovered, though even local custom cannot subvert Torah law, which leads to holiness.

For example, it is custom on the first day of class to pass out a syllabus— a road map, so to speak, of the route of the course. It is a document that assigns means and ends: the syllabus contains objectives to be met, and it sets out the particular path chosen by the particular professor. The syllabus announces what texts will be employed, what assignments are due for each class meeting, the various personal *minhagim* of the particular instructor (attendance preferences, notebooks, etc.), ideological statements explaining the course and the instructor and, of course, procedures for grading. What role does custom play in the development of the syllabus? The document itself derives from custom: let me wonder about this.

I think that this semester I will leave the syllabus blank. For thirty years, I have struggled in the preparation of these road maps. I don't know what is for dinner this evening, but I am required to define exactly what must be read and known six months hence. The syllabus is the imprint of my authority. It both frightens me and protects me. I will say to students when I return to the university classroom: "At the conclusion of this semester, turn in to me a completed syllabus. Tell me where you have walked and what you have seen. Report to me what you have done and tell me what you have learned."

Everything is in accordance with local custom. But how can I be a teacher if all is to be followed according to the way it has always been followed? Learning should effect some change, but adherence to local custom suggests lack of movement. The teacher must listen carefully to the local custom of those who come from disparate towns, and the teacher must listen carefully to the local custom of the school. And the teacher must listen carefully to her own local customs. And then the teacher must study to what holy ends these local customs would lead. Torah law supersedes all local custom, thank goodness. I am holy you should be holy—the standard is relational and not possessive. The local custom is actually not local, I think. Finally, the teacher teaches not curricula nor syllabi, but students.

I think I have gone very far afield. It is permissible. I am, after all, on sabbatical when I am forbidden to work in my fields.

NOTES

1. Again, I am using the English transliteration for the Hebrew word for the seventh day.
2. Schemettah is the Hebrew word for the sabbatical year.
3. I am reminded of the tension between John and Abigail Adams and their respective places in history texts. While the former helped draft and pass the Declaration of Independence the latter maintained the farm in Braintree, Massachusetts, which permitted John Adams to continue his work in Philadelphia. And whose work is the greater?
4. *Minhag* is the Hebrew term for custom, and for purposes here, it refers to local custom. *Minhagim*, below, is the plural form of the noun.

CHAPTER 6

THE SIXTH SAPLING

I love the form argument takes in Talmud. Unlike in the classroom where the right response is inevitably sought, the Rabbis seem rather to eschew answer to pursue method. In the midst of the discussion begun in the previous chapter concerning the length of the workday and the content of the worker's meal, the issue of local custom is suddenly dropped and no further talk about this topic will be undertaken for a considerable amount of time. Oh, the Rabbis will certainly return to the issue, but not until after a remarkable, lengthy and, for my purposes, fascinating and illuminating digression. Specifically, in the midst of this conversation regarding the length of the workday, this remarkable shift occurs:

> Rabbi Zera expounded (and some say Rav Yosef taught): "What is it that is written: 'You make darkness, and it is night, wherein all the beasts of the forest creep forth?'"

A straightforward question, I'd say. Rabbi Zera (and some say Rav Yosef taught) addresses the same psalm as had been cited as proof text for the measure of the workday, but he makes query about the verse (20), which just precedes the verses already made meaningful by the Rabbis' discussion. Now, I have been taught by Western logic that the anonymous editor who at this point introduces Zera into the discussion does so in order to further explicate and clarify the discussion concerning the length of the working day. But this is not the case at all! Zera intends to present an altogether new context and to offer a completely independent agenda for the words of the psalm: he is not at this time in the least interested in the length of the workday. He has, as it were, a different syllabus to pursue.

I am intrigued by the Rabbis' acceptance of the introduction of Zera's radical departure from the sense of their conversation. No one demands that the topic presently under discussion continue to be addressed. Indeed, as we will see, the Rabbis move effortlessly along with this radical shift. I think we are much less tolerant in schools. For me, one of the great scenes in American literature is in Salinger's *Catcher in the Rye.* Holden Caulfield has taken refuge at the apartment of his former English teacher, Mr. Antolini. Before having to return home to admit that he has been thrown out of yet another school, Holden begs a couch for the night in Mr. Antolini's New York City apartment. In the ensuing conversation, Antolini wonders if Holden is at least passing English, and Holden admits that alas, he has, indeed, failed Oral Expression. When Antolini asks why Holden couldn't succeed there, Holden says that he couldn't stand participating in a class that refused to permit digression. "I guess I don't like it when somebody sticks to the point all the time." Holden, I think, would have appreciated the practice of the Rabbis of the Talmud.

Ever the teacher, however, Mr. Antolini, suggests that perhaps there is a time and place for everything and that a topic once chosen should not be suddenly abandoned for one more immediately interesting. I do not believe that Mr. Antolini has recently talked to a group of kindergartners. Antolini prioritizes consistency and derogates contingency; kindergartners, at least, rely on spontaneity and possibility. To them, each statement is equally exciting and derives from equal joy. But logic and uniformity is the order of schoolrooms occupied by Mr. Antolini and his colleagues, as it must be for the world into which these students must go. Steady measured progress is the standard form of discursive practice. We must arrive at World War II by June! In methods and classroom management classes, teachers are taught how to deal with digression, even as students are surreptitiously appointed—and honored—by their colleagues to lead teachers astray either to enliven deadly academic discussions, or to delay the execution of the final test. Digression is anathema to the classroom.

But I am enthralled at the introduction of this dramatic shift in the Talmudic conversation. It is clear that someone desires to change the register of this entire conversation, and I am prepared to follow. Zera wonders what is meant by the earlier verse and then proceeds to answer his own query. I take the liberty of adding medial verse 21.

[20] You make darkness and it is night, in which every forest beast stirs. [21]The young lions roar after their prey and to seek their food from God. [22]The sun rises and they are gathered in, and in their dens they crouch. [23]Man goes forth to his work and to his labor until evening.

Very clearly the distinction between day and night is the subject of these verses. Then what meaning is added by verse 20 that inspires Zera's question? Darkness is the moment of night; night is the domain of beasts of prey. Certainly, the text authorizes as part of the natural order the deadly young lions' hunting. The text give voice to the Rabbis' imperative that the workday must end before the lions come out at night. Indeed, if the worker is obliged to go off to work when the young lions are roaring after their prey "to seek their food from God," then the worker is certainly endangered by the roaring and foraging of these voracious beasts. This peril is clearly not part of the divine plan. It would seem that according to the Rabbis' discussion concerning the length of the workday, it is not intended that one's life should be endangered as a result of the efforts of the workday, which has itself been established to maintain life: "Six days shall you labor"; "For six years shall you sow your field and for six years you may prune your vineyard; and you may gather in its crop." I appreciate this much: not an inconsiderable amount of labor history concerns the struggle over this very issue of safety in the workplace. Schools, places of work for both faculty and students ought, too, to be safe. What safeguards are established to protect the children who must attend these institutions? How do we protect our children from the lions who seek their food from God? I gaze out my window in the early morning dark and I see the strobe lights flashing atop the school buses. There are children waiting at bus stops in the moonless morning for their long journey to school. I think some will return in the dark as well.

I cannot stop thinking about Charlie. He was in Emma's second grade class. He missed very few days of school, though it was apparent to all that school was very difficult for him. Charlie spent many hours in the principal's office, or in forced isolation at the back of the room. Once, he was even secluded behind a makeshift barrier or wall erected to effect some separation between his antics and the business of the classroom. He was at times completely uncontrollable and highly disruptive. When punished, he would often cry. I know only glimpses of his life, but we all suspected that for Charlie, school was one of his safe havens. However, he demanded to be recognized there: it was daytime and Charlie had gone out to work. But the lions ate him, I think.

✡

Rather than deal with mundane issues concerning the length of the working day or the character of the luncheon meal or snack, Rabbi Zera offers a profoundly metaphorical and clearly individual interpretation of these same lines. Zera expounds:

"You make darkness and it is night" refers to this world "which is like night." "Wherein all the beasts of the forest creep forth"—these are the wicked in it, who are like the beasts of the forest. "The sun rises, they gather themselves together and lay themselves down in their dens." "The sun rises"—for the righteous. "They gather themselves together"—the wicked [are gathered] to Gehinnom. And lay themselves down in their dens"—there is not one righteous person who does not have a dwelling place befitting his honor."

Well, here is an interpretive leap and a philosophical agenda that startles me awake! This Rabbi Zera, by this radically interpretive use of text (though some say it derives from the teaching of Yosef), certainly intends an advocacy of a particular worldview. I, at least, and with all my supposed training, can discern no apparent context in Psalm 104, and certainly not in the context of the conversation about local custom and the length of the workday, for such a thematic development as Zera (and some say Yosef) proposes. The introduction of this line of thoroughly digressive interpretation suggests that Zera (and some say Yosef), or even the anonymous discussant who has offered this teaching, would prefer to pursue from this text some other speculative direction than that of prosaic labor issues. In his exposition, Zera (and some say Yosef) means to connect the physical—darkness and light—to the metaphysical, this iniquitous world and the promise (or threat) of the next. I think he means to connect issues concerning the legal—the responsibilities of the employer to the employee—with those pertaining to the ethical—the rewards of righteousness. Zera (and some say Yosef) is concerned not with the physical darkness of night, but the spiritual darkness of wickedness. Zera (and some say Yosef) is concerned not with the physical brightness of the sun, but the spiritual brightness of the righteous.

Neither Zera nor Yosef say *why* this world may be likened to night, but perhaps we might guess that it is because the wicked roam through it. If this is so, then we are all potentially prey to the wicked who may use the physical darkness to cloak themselves and their deeds. We are all prey to the beasts who stalk their prey at night, and we are all are vulnerable to the wicked who prefer the darkness of night. Who darkens the daylight and makes it night? Who might they be? And aren't the righteous those who study? Aren't the righteous those whose work is informed by their study?

And then having started out on this hermeneutical adventure, Zera (and some say Yosef) proceeds to address the very same lines, Psalm 104:22–23, which the Rabbis have cited in determining the length of the workday. Perhaps Zera assumes prerogative to do so because the lines in the Psalm refer, as does he in his explication, to a going out and coming in. Zera's interpretation, then, creates its own context. And having taken his interpretive leap, Zera lands not on issues concerning the length of the workday, but on the metaphorical day of sunlight that shines only for the righteous.

Zera continues his interpretation of Psalms 104:22:

> "The sun rises"—for the righteous. "They gather themselves together"— the wicked [are gathered] to Gehinnom.[1] "And lay themselves down in their dens"—there is not one righteous person who does not have a dwelling place befitting his honor. "Man goes out to his work"—the righteous will go out to receive their reward. "And to his labor until the evening"— concerning someone who has completed his work until the evening.

Rabbi Zera says (and some say Rav Yosef taught) that it is only for the righteous that the sun will rise in the world to come—only they that will receive their reward at their death. In Zera's interpretation, the righteous, those who walk in sunlight, go out to their work, now defined as going out into their reward. The righteous engage not in labor, but in love. As a result, the righteous will flourish in their efforts with great joy, or they are assured of a place in the world to come as a result of their work in this world. As long, that is, as that work continues until the evening—until their death.

Perhaps Zera introduces his perspective because he means to make some connection between the Mishnah concerning practical affairs—issues perhaps of ethics—and metaphysical matters—issues of righteousness and wickedness. The worker has become the righteous and the length of the day the years of a person's life. Work has become reward and righteousness defined as the daily effort. Perhaps for Zera there is a relationship between metaphysics and ethics—the entrance into the next world and the stance in the present one. Perhaps, Zera suggests, there is, in fact, no metaphysics, there is only ethics: work *qua* righteousness. The connection between work in this world, the daily labor addressed by the Rabbis, and the work that leads to the next—the achievement of righteousness—dissolves the gap between metaphysics and ethics by insisting that the next world is contingent upon activity in this one. In this world, the Rabbis suggest, there are the wicked and the righteous. I am postmodern enough to be wary of easy dualisms; indeed, the intricate arguments in which the Rabbis of the Talmud engage—for example, about the power of local custom— bespeaks their own doubts concerning certainty. It is, I suspect, we who must decide who is wicked and who is righteous. The Rabbis suggest that this is an issue of work and relationships.

✡

I like this Rabbi Zera. I mean, he is listening attentively to the conversation at hand, but he is clearly not interested in its present level. I am reminded of so many moments in the classroom: There I might be, teaching a particular novel or short story or poem, and the Socratic discussion proceeds in a manner of which Socrates might approve, which is to say, one led by me in a certain direction to fulfill the objectives set by my curriculum! This was an event in which I knew all of the answers to the questions I knew I was going to ask—or at least, I pretended that I knew the answers. This was a moment when I believed that a book's meaning was so important that everyone should get it. It was a time when I actually believed that literature would save these children from . . . well, from my own difficult childhood when literature served as a safe haven and sheltered me. And suddenly, in the midst of discussion in which I was thoroughly engrossed, someone would make an unexpected and undesired response instantly changing the direction and register of the conversation and the interpretation in which I had invested so much of my time and energy and intellect. And I would respond snappishly to the student and assert the centrality of my direction. I had been sanctioned by my training to ignore or even dismiss Rabbi Zera, I am afraid to say. But sometimes . . .

Once, during a particularly bleak moment in New York City history when the numbers of homeless people increased precipitously as a result of some bureaucratic ineptness and cruelty, I was teaching Tennessee Williams's *A Streetcar Named Desire*. This was a standard eleventh grade text—standard that is, for the "regular" and the honors classes. The basic classes, regardless of age and/or sophistication, were instructed in (that is the appropriate term) the rules of English grammar and were offered texts of less complexity—say, Zindel's *The Pigman* or Richard Peck's *A Day No Pig Would Die*. But, I would not let these students be treated so. I had previously read aloud to them every page of Salinger's *Catcher in the Rye* and had now turned in Williams's play to a portrait of yet another lost soul. I decided to teach Blanche DuBois as a bag woman, as one of those unfortunates who had been cruelly cast out on the streets of New York City at this particular historical moment in the City's history. They roamed about homeless and forlorn, standing on the corners with their suitcases containing all of their earthly possessions waiting, as it were, for the streetcar.

My department chair, who originated from Kentucky, asked me during a post-observation conference, if I thought that as a result of my pedagogical decision the students would still appreciate Blanche as a decayed flower of the southern aristocracy. (I wish I had been studying Talmud then—I would have asked, "From where do we learn this?"). I don't remember my response exactly, but in substance I agreed that as a result of my pedagogical decision, Blanche might not be understood as a fallen aristocrat. However, I offered, perhaps the students might gain some insight into the present moment in our history from this particular reading of *A Streetcar Named Desire*. Terry Eagleton once said that the best performance of a text is one that sends you back to the original. Rabbi Zera is a master of this prod.

✡

Once Rabbi Zera has opened the hermeneutic trunk, the other Rabbis reach willingly in and draw out their own interpretations. A tale is told:

> Rabbi Elazar the son of Rabbi Shimon met an officer who was arresting thieves. He said to him,
> "How are you able [to detect] them. Are they not compared to beasts, as it is written: "Wherein all the beasts of the forest creep forth"?
> (There are some who say that he quoted the following verse to him: "He lies in wait secretly like a lion in his den.")
> "Perhaps you take the righteous and leave the wicked?"
> He said to him: "But what can I do? It is an order of the king."
> He said: "Come and I will teach you what you should do. Go at the fourth hour to a tavern. If you see a man drinking wine and holding a glass in his hand and dozing, ask about him. If he is a rabbinic scholar and is dozing, he rose early for his studies. If he is a worker, he rose early [and] did his work. And if his work is [done] at night, he was stretching [metal wires]. And if [he is] not [one of these], he is a thief, so arrest him.

And so it happens that from the textual interpretation of Psalms and the discussion of the length of a workday, we move to a certainly practical but certainly extraordinary discussion regarding human behaviors. Elazar is giving instruction to the officer of the king—to the enemy—regarding the detection of thieves. Rav Zera's hermeneutical chain that links the beasts that roam at night to evil people who roam at night and which clasps together the physical to the metaphysical, is now linked by theme and contiguity to Elazar's desire—and method—to rid the world of the evil people. Of course, it is important to rid the world of such evil people that the righteous might go to reward in the evening at the end of their work. That reward seems to include at least a safe passage home. Elazar contends that the righteous, carrying their honestly earned wages in their hands, ought not to fear mugging on their return home. More, Elazar would rid the world of the iniquitous and leave it to the righteous. He approaches an officer of the king whose responsibility it is to arrest those suspected of criminal behavior. Elazar is concerned that the soldier of the king not arrest the righteous with the wicked: how, indeed, might the latter be recognized?

Elazar asks the king's officer: How are you able to detect them, the wicked? Aren't they, Elazar considers, like the beasts of the field, which come out only at night when detection is difficult in the darkness and when no one else is about? And if the wicked are so carefully concealed in the world, mightn't we be in danger of arresting the righteous and leaving the wicked free to roam? By physical appearance alone, the wicked and the righteous are indistinguishable. Indeed, the soldier admits to the difficulty of his job—admitting that he really doesn't know what he is doing. The soldier of the king has little confidence in his methods; indeed, he does not seem to have method. "What can I do?" he says. "It is my job."

This rationalization moves through history in so many forms—"I was just following orders;" "This is going to hurt me more than it hurts you;" "Do what I say and not what I do!" "I have to punish you for your own good." "For the sake of the children, I must teach to the test that measures externally defined standards." I think this is always a cowardly and contemptible explanation. It defines nothing but the absence of responsibility. Elazar rejects the soldier's rationalization.

Rather, Elazar proposes a method by which the wicked and the righteous might be distinguished. And his method concludes that it is by activity that the wicked and the righteous might be known. Neither by God's grace nor arbitrary election might the righteous be marked; unlike the teachings of Western culture, there is no physical mark or appearance that distinguishes the wicked from the righteous; that distinction arises as a result of action and not aspect. The tavern's customers at the fourth hour are either students exhausted from studying, or they are workers weary from labor. Or they are thieves! Work and study are sanctioned activities in the world. Engagement in that work legitimately exhausts; when the students or workers pause for respite and relief, they necessarily fall asleep.

Now, we may know the substance of the effort in which students and workers have been engaged, but as for the others in the tavern, what might s/he have been doing that would lead to such exhaustion? The answer seems obvious: they have been stealing, and they must be thieves. Or perhaps we might consider the activity that the tavern-goer has *not* been doing as equally important: s/he has been neither studying nor working. In which case, the individual is yet a thief, and s/he has been stealing time.

I am wondering if Elazar hasn't accurately described my personal world. When else should one be discovered dozing in the tavern but in the respite after study or work, to both of which we must return anon? And as for the others resting by me—well, what else must they be if they be neither students nor workers. They must be thieves. The thief has no place to go except the tavern: he would neither study nor work, and so in what activity but sleep would he engage having arrived from his exertions in the tavern. And besides, as Zera has said, it is only the righteous who have homes to which to retire, and only students and workers are righteous. There is no ease at home for the wicked, for the wicked have no home.

I tell my students: "I know that you have social lives. For this semester your social life should consist in meeting your friends to talk about this course." I am not being facetious when I say this, though people often laugh nervously: I think that class ought to be both direction for and break from solitary study. Once embarked on serious study, why would one break from that exploration. What could be more appealing than study? As we are educators, then our lives are dedicated to the study to which we rise early and stay late. Not unlike the description of the worker's day, in fact. We are, indeed, workers. We pause briefly at the tavern as a respite, so to speak, before our return to our books or our home. Where are these books to which we go?

Unlike the Mishnah with which I started, this story about Elazar suggests that local custom may not be used to distinguish the righteous from the wicked—the wicked are indeed wicked, and they might be readily discovered. Evil is neither relative nor local: it is absolute.

I stand before a classroom filled with people. Which are the righteous students and which the wicked?

There is work and there is study. Study and work are equated here— if one is not studying then one is working; if one is not studying, one should be working. Both activities, however, require rest. All work ceases, of course, on Shabbat. And every seventh year, the land shall receive a Sabbath rest—the schmettah year. This reminds me of my earlier question concerning the length of the students' workday—I too often ask students to come early and to stay late—and here we have a response over the span of almost 1500 years. Yes, scholars awaken early and stay late, and it is for this reason that they fall asleep at their meals and even at their cavortings. I wonder: did those students asleep at the rear of my classroom awaken early?

But if one is engaged in neither study nor work, then one is a thief.

I have always worked in the classroom. I have been a teacher for almost thirty years—certainly more than half of my life. If I include the times of my own education, I have spent my entire life in classrooms. I have spent time in the taverns as well, even in the metaphorical kind. I wonder now, if the officer found me there, how would I be recognized?

I adore being out and about in the world. I think that many of these journeys have been occasions of study: I have transformed even the bacchanalia of the Grateful Dead into matters addressing pedagogy. I have, I surmise, quoted Bob Dylan as often as John Dewey. I have spent thousands of hours in the movies. But I wonder now how many of my activities have been distractions from the more serious occupations of study and work? I wonder what Elazar would think of the effort I make on my sabbatical. And this work to which Elazar refers: would it also include the raising of my children? There must be no sabbatical for those fields. Indeed, I find sometimes that I must go to the tavern as a result of this effort. Teachers go regularly to the taverns from their days in school. And there they bemoan their lives in school.

In a graduate class I taught during the fall of 1998, I observed an interesting phenomenon. The class citizenry consisted of folk already engaged in, or committed to, the teaching profession, and we gathered late afternoons and early evenings every week to study. And many study sessions began with vigorous complaint—about students and parents and administrators and colleagues—about the system and the government and the standardized tests—about poverty and disadvantaged homes and feelings of powerlessness. And suddenly LaMont said, "You know, sometimes I sit here and I listen to you folks and I don't know if I want to teach at all." I was startled: it struck me that if, indeed, life was as horrible as these teachers described, then who, indeed, would choose to be a teacher.

"Sometimes," the Rabbi said, "I am so disgusted with the world that I would like to vomit it up. But then I think, is the world mine to throw up?" Well, that is one answer. Perhaps, I considered, it was necessary to first vomit up all of the bile and the daily difficulties and indignities before we could begin to produce new ideas. Perhaps the vituperative talk was a cleansing—like the mikvah, the ritual baths of Jewish custom. One assumes a certain impurity before entering, and after immersion there one feels cleansed. But how sullied the water must get after every bather! The Rabbis mandate that the mikvah consist of swirling waters to ensure the hope of renewal. Is it my responsibility to ensure that the waters swirl?

Slowly, almost imperceptibly, the conversation altered tone, and the talk turned to repair—given the conditions in the world, what could we teachers do to heal? Perhaps LaMont's comment confronted them with their spoken misery and they were humbled. Perhaps they remembered what it was that they meant to do every day in their classrooms because it was to this purpose that each of them individually addressed themselves for the remainder of our evenings. Finally, it is *tikkun olam* that these teachers believed to be their purpose. It is a familiar theme in Judaism— *tikkun olam*, the healing of the world. It is a complex notion that different ideologies of Judaism interpret in a variety of ways. What each interpretation shares, however, is that the work of this world is its repair. In our classroom that late afternoon, the voices turned soft and reflective, and each talked of the work they had accomplished that day. It was never of standards, or performance, or curriculum duties that each spoke; rather, it was to the human touch that had been made. We talked, I believe, of healing.

Another story: It was Yom Kippur, the Day of Atonement, and all of the congregants waited anxiously for the Rabbi to arrive that the Kol Nidre prayer might begin. One woman sitting in the congregation glanced nervously about for she had left her child asleep in the crib, but as the minutes passed and the Rabbi did not arrive she worried that the baby might awaken and be alone. Finally, after some minutes and the Rabbi still had not appeared, the woman thought to herself, "I guess it will be some time before they begin . . . I'll just run home and look after my child to make sure it has not yet awakened. It will only take a few minutes." And the women ran home. The house was yet quiet. She turned the knob of her front door and quietly put her head into her child's room. There stood the rabbi holding her child in his arms. He had heard the child crying on his way to the shul and he had cradled and played and sung until the baby had fallen back asleep (Buber, 1947).

In our schools and in our classrooms, I hope we do not miss the weak cry of the child for the blare of the curriculum.

✡

Elazar has offered what he takes to be a foolproof method for discovering the wicked. Ah, for the comforts of methodological certainties. In so many classrooms, teaching is only about absolute methods: the pedagogy of the how-to. Students tell me, "Please, just tell me how to do this and I will do it. I will do it well. Just please, please don't trouble me with ideas." Methods proliferate. Methods classes proliferate. Teaching has been transformed into a set of directions not unlike those I cannot follow when putting together my children's toys, or my furniture, or anything else that comes in a carton with directions written by someone skilled in writing directions. Peruse any methods book and discover the myriad numbered lists of how-tos, or read the plethora of acronyms referring to methods and techniques. Teaching is almost never about studying, but about how to study, not about learning but about achieving, not about healing, but about administering and assessing.

✡

I've never written this way before—I am following some idea that began in Bava Metzia and I have no idea where it (or is "it" a "they"?) might lead me. Here, today, there is a great deal of underbrush and no discernible path that I can readily follow. Yet, in this shemettah, I am forbidden to sow or to prune nor to reap nor pick. Just keep writing through the thicket.

Elazar says that I must rise early to my study or my work if I am to earn that drink at the tavern. Else I am a thief.

✡

Perhaps Elazar refers here to the responsibility of the teacher to recognize students. Here I refer not only to the Levinasian face-to-face look that is the ethical, ontological moment, but also to the acknowledgment of the quality of effort made by a particular person—the business of evaluation. My judgment of their endeavors finally identifies those before me as either students or thieves. Here is the very notion of assessment that is the subject of voluminous academic publications. Evaluation: how-to. Finally, assessment—or evaluation—is the ultimate source of the teacher's power. Of my power. All communication between myself and students—even that ontological face-to-face ethical moment—is mediated by the understanding that at the course's end it will me who will give the grade that defines the student by summarizing the particular effort or work each has been thought to make. It is finally the teacher—me—who must recognize the student. I am the government's officer. Elazar says that if that dozing patron is a student, then it is because he has risen early for his studies. Elazar believes that students study. Elazar knows that the function of the student is to be engaged in study and that when the student is not so engaged, then he is napping at the tavern before returning to his studies.

There are many students in my classroom. Some of them are asleep.

I would know which is the student and which is the thief who steals from me my time! Or perhaps it is I who is the thief?

✡

Elazar's method of discovering thieves was reported in the house of the king, and he was summoned to the palace, antedating, I suppose, the old adage that if you build a better mousetrap the world will beat a path to your door. Since it is his own idea, the king proclaims, then let Elazar accomplish it.

> Let the one who reads the letter be the messenger.
> They brought Rabbi Elazar the son of Rabbi Shimon, and he would arrest thieves.
> Rabbi Yehousuah ben Korhah sent [this message] to him: "Vinegar, son of wine, how long will you hand over the people of our God for execution.
> He sent [back] to him: I am destroying thorns from the vineyard.
> He sent back to him: Let the owner of the vineyard come and destroy his thorns.

Elazar has been hired by the government to arrest thieves, though not without some protest from his community.

There are responsibilities, it would seem, to the production of ideas. The king has hired Elazar, whose method for the detection and arrest of thieves seems to the monarch sound. The King for one (the officer for another, I would imagine) is pleased. But others, Rabbi Yehousuah ben Korhah for one, doubt the virtue of Elazar's efforts. If Adonai wishes the garden cleaned, he argues, then let Adonai do the work. Elazar defends himself saying that he is enhancing the beauty of the garden by removing from it the thorns, but Rabbi Yehoushuah counters by telling Elazar that it is not a human but God's concern to care for the metaphorical garden. Let the owner of the garden remove the offending thorns, Yehoushuah admonishes Elazar. God's garden should be left to God's care and God's position (as gardener? What about kindergartens?) ought not be usurped by human beings aspiring to godlike judgment. I am sensitive to this conflict. Who should assume the responsibility for identifying the student from the thief? I am often called a gatekeeper: it is my function to admit some to further study and to prevent others from such progress. Frost's narrator says, "Something there is that doesn't love a wall." Or, in my case, it is a gate. Whose garden am I protecting, and against whose entrance am I on guard? There are some who I have kept out—I found them sleeping at the tavern during the fourth hour. I named them thieves. Were those I kept out stealing time? Against what is the gatekeeper to guard? If I am always the student, then how can I be the gatekeeper as well? I am a teacher, but always first I am a student. Does someone guard the gate at which I call? Perhaps it is the editors of journals and publishers who return my manuscripts. Ironically, I think them thieves.

It is not the number of hours spent in the classroom which ought to be quantifiable; it is the devotion those hours entail that is important. How will we evaluate that devotion? How assign a grade? What would that grade mean? Who should ultimately be responsible for that grade? Who should be responsible for removing the thorns from the garden? By what standard do we measure the future teacher? Is it grades or character we value? And if the latter, how should it be measured except by the engagement in study that is unending and for which, the Rabbis say, there can be no measure. Could constant study require a summative evaluation? I think student portfolios are an attempt to deal with this issue: but what would a portfolio that reflects not product but process look like? Especially when its product is all method? How will a student represent his/her devotion? How do I reflect my own? Is this sabbatical a reward for that fealty? Thoreau once said that we would always rather be living life than writing about it, but I wonder if this was true of Marcel Proust?

For myself, I prefer being with my children than photographing them. If I am always engaged in study, then what do I put in my portfolio? Study must acknowledge the world, though it need not produce for it. In the Yeshivah, there are no grades; there is simply the acknowledgment of sharp and dull minds. Let those who feel unwilling to dedicate their lives to study remove themselves from the academic garden. Such self-selection would obviate my role as gatekeeper; but then what would become my role in the garden? My children play in this garden. Do I own the garden? I am also a worker in the garden. Ah, but this is the subject with which this discussion began: the consideration of the hours of work and the place of local custom in the definition of my due and my responsibilities.

✡

This story of Elazar's method of discovering thieves reminds me of another responsibility the teacher accepts: evaluation. At the end of each quarter or semester, teachers tote up the various numbered or lettered grades in their grading books and assign an evaluative measure to the student. At the end of each quarter or semester, teachers define who is a student and who is not, who shall live and who shall die, who shall perish by fire and who by water, who by hunger and who by thirst, who will live in harmony and who will be harried, who will enjoy tranquility and who will suffer, who will be impoverished and who will be enriched, who will be degraded and who will be exalted. It is an awesome responsibility that I suppose accounts for the huge salaries granted to teachers in our public schools.

The Gemara continues:

> One day a certain laundryman met him [Elazar] and called him "Vinegar, son of wine." He [Elazar] said: "Since he is so impudent, infer from this that he is a wicked man." He said to them: "Arrest him," [and] they arrested him. After his mind was calmed, he [Elazar] went after him to redeem him, but he was unable [to do so]. He said of him: "He who guards his mouth and his tongue guards his soul from troubles." They hanged him. He stood under the gallows and wept. They said to him: "Master, let it not be grievous in your sight, for he and his son had intercourse with a betrothed maiden on the Day of Atonement." He laid his hand on his belly [and] said: "Rejoice, my belly, rejoice."
>
> But even so, he was not at rest. They gave him a sleeping potion to drink and brought him into a house of marble and tore open his abdomen. They removed from him basketsful of fat and placed [them] in the sun in Tammus and Av, but it did not decay.
>
> But no fat decays!
>
> No fat decays, [but if it contains] red streaks, [it] does decay. Here, even though there were red streaks, it did not decay. He said of himself: "My flesh too will dwell in safety."

This is an interesting, though admittedly, bizarre story. Elazar, whose method for distinguishing the wicked from the righteous so appealed to the king, had been granted by order of that king license to employ his methods. However, the story above recounts how Elazar had caused the arrest of someone not for being in the tavern at the wrong hour, so to speak, but for personally offending him: the man had called Elazar a name, "Vinegar, son of wine." Indeed, there is even some sense in the story that the aspersion cast upon him refers to Elazar's newly appointed position as a government official arresting fellow Jews. Once power is proffered, as we well know—even of that to arrest—it is not suffered lightly or without condemnation by those subject to it. Nevertheless, such proffered power may be exercised with relative impunity. Elazar, offended by a certain laundryman, took recourse made available to him by the law to redress the calumny he has personally suffered. The arrested man is condemned to death.

But Elazar developed second thoughts. Despite assurances from others of the man's wickedness, Elazar, having had time to reflect upon the incident, perhaps even honest enough to appreciate the role his personal motives may have played in his arresting order, expresses doubt regarding the man's guilt and pleads for his release. But when the king refuses to relent, Elazar quickly acquiesces and consoles himself by quoting Proverbs 21:23 as justification for the accused man's fate: "One who guards his mouth and his tongue guards his soul from troubles." The man was careless with his tongue and now he suffers justifiably for his reckless abandon. Elazar, we are coming to appreciate, demands a strict and unwavering standard of behavior. Perhaps Elazar assumes that a student engaged in constant Torah study will behave with the same righteousness that derives from and leads to that study; Elazer expects, as well, that an assiduous worker would also have neither time nor inclination to engage in such vituperation. Elazar justifies his evaluation by whatever means necessary. I think Elazar imagines a more just world than the one in which I live.

We don't always like our students, nor do they always speak well of us. The official power by which we are licensed permits us to condemn not a few. I do not doubt that I have sentenced several because they referred to me as "Vinegar, Son of Wine." Of course, Elazar assures me, if they had guarded their tongues, then no judgment would have fallen upon them. On the other hand, the verse in Proverbs suggests that when I do not guard my tongue, when I insist that I am right, I define my evil. Proverbs suggests that the good do not boast; they are too busy studying. If this is so, then my evaluation of students is not only superfluous, but vain. The trust the wise inspire comes not from their words, but from their restraint. It is not that the wise are silent, but that they are humble. As study is interminable, then any moment is inappropriate to make an ultimate interpretation or judgment; the wise refrain from completion. (Is this a quietism that study promotes? Is this the immorality of reading?) As study is interminable, there is never a moment when grades are appropriate; *the wise refrain from completion.* But the sluggard who is incapable of disciplining himself or practicing restraint is subject to his desires. He is easily distracted and is incapable of either study or work. He is a thief. Does he steal the value of life? The thief requires judgment. Elazar suggests that arrest is appropriate; in schools, the death sentence arrives in the form of grades.

The grades that we assign are, indeed, as permanent as any death sentence. Lisa Haron, a ninth grader I taught many years ago, but whom I cannot yet forget, came to me sobbing. "Lisa," I asked, "what's the matter?" "I got a 'C' on this paper," she said through her tears. Indeed, she had; I had given her that 'C.' "But, Lisa," I responded, "its only October; your grades will change as your writing improves." What must I have been thinking about back then? Was it Lisa who had earned that C, or was it her previous teachers to whom that grade should have been assigned? Was it I who had so poorly prepared her to complete the assignment? What could that 'C' have meant so early in the school year of her first year in high school? That Lisa had much to learn? Didn't I already know that? Obviously Lisa didn't know that yet—she was concerned with some vague distance goal. She wailed, "But how will I get into law school?"

Rather than worrying about the sentence the grade had rendered, Lisa could have been learning the beautiful endlessness of study. She was, perhaps, distracted from learning by the evaluation of it. Wasn't that 'C' a *non sequitur*? It interjected absurdly in a situation to which it added nothing. What could that grade announce about Lisa's anticipated learning? If everyone knew that Lisa was in school to learn, then what purpose was served to assign her that 'C' so early in the year? Or ever? What could that grade have meant?

What does any grade mean? Grades are only meaningful as a means of comparison. My 'C' exists only when compared to his 'A' and to her 'D'. Grades are founded in competition and of themselves say nothing at all. And if everyone receives an 'A' because they learn, then what is the standard that is measured by that 'A'? I was hoisted on my own petard. And so I joked, "Lisa, if this 'C' is a hindrance to admittance to law school you come back and I'll change it." But I think what I actually communicated to Lisa was that the 'C' I had assigned her contained no meaning. At least for me. She, on the other hand, was devastated by it. I felt like "Vinegar, son of Wine."

Elazar does not insist upon his correctness and is not therefore, evil; rather, Elazar responds humanly—with anger at the slight—and then regrets his impulsiveness. But the powers render his contrition ineffective. He is too late.

And this disquisition on evaluation begins to address the significance of the growing plague of scores found on student records from the standardized tests that, like cancer cells, multiply uncontrollably.

I am always disturbed by grades. I dread the end of the term when I am responsible for grading. My stomach begins to hurt. Indeed, isn't Elazar's stomach exactly his problem and isn't its affliction tied to his summative evaluation of the laundryman? It is my responsibility to measure up a student's semester with a single letter grade. I am overwhelmed by the act: on any single day every student earns an A and on any single day every student earns an F. Which day shall I evaluate? Are we all merely the sum of our averages? I am thoroughly inadequate to the task. Or perhaps evaluation is a meaningless task of which adequacy is no measure. I know what the advocates of standards argue; Diane Ravitch

(1995, 12) writes: "Curriculum standards describe what teachers are supposed to teach and students are expected to learn . . . Performance standards describe what kind of performance represents inadequate, acceptable, or outstanding accomplishment." The Wisconsin Department of Public Instruction (Department of Public Instruction, 2006) announces: "Academic standards specify what students should know and be able to do, what they might be asked to do to give evidence of standards, and how well they must perform. They include content, performance, and proficiency standards." Oh, I know what the standards advocates desire: I just don't believe them. Wisconsin's Department of Public Instruction has proposed revising teacher education programs to a performance-based approach to licensing. Rather than counting on the fact that students in college completed the requisite coursework, it was suggested that there be a measure of performance to substantiate that the candidates for a license had the knowledge and skills needed to succeed.

What a fascinating paradox. The Department of Public Instruction, on the one hand, moves to a performance-based approach to licensing and then, on the other hand, insists that certain knowledge and skills be required in order to be licensed. Those standards will be assessed by a standardized test. It is a closed system where the knowledge base is static. Nothing enters; nothing leaves. Entropy. Standardized tests approved by the state superintendent will be used to assess content knowledge. Well, as soon as we find out what material on the tests is approved, we will know what knowledge and skills are required for licensure, and then those rules and regulations, which are declared by the Department of Public Instruction as the bane of individuals and colleges, will return with a vengeance. Standardized tests are standardly marked—usually and at least, in large part, by some machine. Will the scholar with a 74 be denied license and the scholar with a 75 be so granted? What is the difference between the scholar earning a 74 and the one earning a 75. I wonder if the Lisa Haron to whom I gave a C is a good lawyer and a good person. My grade accounts for neither.

Perhaps there is another way of assessing learning. Thus far I understand Talmud as instituting at least two standards regarding the inefficaciousness of assessment: on the one hand a scholar is constantly engaged in study, save when he is asleep in the tavern. This is Elazar's standard. On the other hand, the scholar is one who takes a specific command and elaborates it to enhance his life; the student interprets and creates custom. But neither standard is liable to immediate evaluation; indeed, it is not the content of their knowledge but their engagement with knowledge that is the measure of assessment. Any judgment regarding content seems grossly inappropriate. How can we begin to consider measures for engagement in study?

Last night Emma came home from ice-skating sobbing. My daughter has been skating for several years now. Every Wednesday during the long Wisconsin winter months, she has taken lessons offered by the recreation department in the university town in which we reside and as a student she has moved steadily and slowly through the various skill levels. She loved to skate. This night she was in tears. "What happened, darling?" I asked as she came running into the room, threw her arms about my waist and burrowed her face into my chest. "I failed. I didn't pass Basic Six. Martha said I didn't even come close to achieving mastery of the things I had to do to pass. I hate Martha." I didn't know what to say. She had called her teacher, "Vinegar, son of Wine"!

But I am being specious: actually, I responded immediately. I said, "Emma, if Martha said what you said that she said, then she is not a very good teacher. No teacher would ever speak so rudely to a student." Of course, that is true, though I am familiar with not a few teachers who speak even more harshly about their students and sometimes even to their faces. But finally, Martha is not a teacher; rather, she knows how to ice skate better than my daughter. I assume Martha has passed at least Basic Six level. She has passed the standards that are set for content area knowledge, I suppose. Of course, I do not think there is a pedagogical component required by the recreation department for those who will teach ice-skating to the children. More is the pity, I say. Martha may know how to skate, but from her behavior it is clear she may know little about children and teaching. But then, except for evidences of truly bizarre behaviors, personalities and peccadilloes are not a legitimate criteria for granting or denying licenses to teachers, and so many are licensed who have not the character for teaching.

Emma does not want to be an Olympic skater. She wants to skate. Martha's designation of her failure got in the way of Emma's skating.

But even that isn't really the issue right now and right here. What are the standards that Martha looked with and for when she judged my daughter's achievements? I assume that there is some level of expertise that marks requisite learning for Basic Six, but I cannot for my life fathom what they might be in *this* program, or how it could be so objectively assessed. In the booklet that marks Emma's progress with little stickers there is simply the statement of the competence, but not the criteria on which performance is judged. Oh, we have watched enough Olympic competitions to understand the point ratings, and we accept almost without question the judgment of the judges. I listen impatiently to the sportscasters sitting safely in their booths point out the imperfections I cannot see in the performance. Finally, I do not care who wins or loses, who will wear the gold and the silver medal. The skaters are for me interchangeable, which is to say, that who wins today will lose tomorrow. What they have in common is that they all return to the practice arena—the study, as it were—the day following the performance. By the standards of sports they are professional because they earn money as a result of their performance, but by my standards they are professional because they are always studying. Of course they improve. And even by Olympic standards, what exactly is the difference between a total score of 46.9 and a 47.1? And does it really matter? Is it really true that a sprinter who runs a hundred-yard dash in 9.9 seconds is qualitatively better than a sprinter who runs it in 10.0 seconds? Why? Wouldn't tomorrow's race yield, perhaps, a different result with the same times? How could my daughter who skated at least twice every week not *come even close* to achieving mastery of the things necessary to pass? How could she have come that far and be yet so ignorant of her inabilities? What can be said of her teachers?

Emma skates once or twice a week. She is yet a child and remains relatively unfocused in her interests and desires. There are some constants: last night I again painted her nails. And to my untrained eye, she skates rather adeptly. She is nine years old and she is not Michelle Kwan. But then, she has no desire or need at this time to be so. She is, rather, competent and content on the ice.

Did Martha wonder if Emma had stolen her time? But perhaps Lisa was correct. How can I know that this meaningless C did not stop Lisa from going to law school? Or that Emma's failure will keep her off the ice again? Or diminish slightly the smile she wears when she skates?

Ah, I wonder: what if our standards are not appropriate ones? What if we are wrong? How, indeed, can we believe so highly in our certainty? "Standards must be rigorous and world-class, reflecting perspectives in Wisconsin, across our nation, and around the world." I truly wonder, what is at stake that we could break the heart of a little girl who loves to ice skate for the achievement of a set of standards? Albert Speer fulfilled and surpassed all the standards set for the study of architecture. Look what he designed! Richard Milhous Nixon satisfied all the standards in his curriculum! Bill Clinton was a Fulbright Scholar! Newt Gingrich fulfilled all the standards necessary to become a teacher! But he wasn't very ethical. None of the above was. Of what service did the standards accomplish?

✡

Elazar finally is absolved of all guilt, when following his operation, it is discovered that though flesh and blood still attach to the fat removed from his body, the fat does not decay even in the hot sun of summer—the months of Tammuz and Av. Elazar, relieved, passes judgment on his innocence quoting Psalms 16:9: "My flesh too will dwell in safety." Despite his moral lapse, his corporeal flesh yet dwells secure. Thus, Elazar declares himself innocent for his condemnation of an innocent man. He must have breathed a heavy sigh of relief—though not as heavily as he might have breathed before all of that fat had been removed. But again, I go to the quoted psalm and I discover an interesting context. The psalm reads: "I will bless Adonai who has advised me, also in the nights my intellect instructs me. I have set Adonai before me always; because he is at my right hand I shall not falter. For this reason my heart rejoices and my soul is elated; *my flesh, too, rests in confidence.*" On the one hand, the psalm acknowledges dependency—I will bless Adonai who has advised me—suggesting that we are not free to act as if we act alone. But on the other hand, the psalm expresses that though the study during the day derives from God's word, the intellect in the night—based I assume in the day's study—plays a part in the choices of our action. An interesting idea: while we are engaged in study, we are somewhat free from connections to activity in the world; we are, as it were, enclosed in our caves, in our studies, in our basements. But at night, when study is finally over, then our intellect instructs. I wonder if the psalm refers to the efficacy of dreams? Dreams, as might any object, be used.

I had a dream. In it I was attempting to get home. It did not seem difficult. But I did not make it.

✡

And also to Rabbi Yishmael the son of Rabbi Yose such a thing occurred. Elijah met him [and] said to him: "How long will you hand over the people of our God for execution?" He said to him: "What can I do? It is an order of the king." He said to him: "Your father fled to Asia; you can flee to Laodicea."

There are other ways to deal with the imposition of standards. There is resistance. When chosen as the monarch's instrument of justice, Elazar accepts the position; so, apparently, did Rabbi Yishmael, the son of Rabbi Yose. But the prophet Elijah offers an alternative to following orders. The prophet, Elijah, confronts Rabbi Yishmael: "How long will you hand over the people for execution?" And when Yishmael rationalizes his complicity with loyalty and duty, Elijah reminds him that his father had more principle: your father, Rabbi Yose, refused a similar position by fleeing to Asia. You, Yishmael, can flee to Laodicea.

It has often crossed my mind to cease being a teacher. But I have never acted upon the thought. And every semester, I hand over some people for execution. Over the years, however, I hope it has been less and less. Levinas says that it is the Other that ensures the existence of the world. Without the Other, everything is only me. When I destroy one person I destroy the world.

I dread the end of the semester. I dread standards.

✡

We live by stories. I do not mean this as an idealist position—I am Marxian enough to know that we live by bread. We eat so that we might continue to live, to reproduce ourselves, as it were. But I believe as well that though we physically live by bread alone, the manner of our eating plays an integral role in creating the story of our lives; we become, as it were, not only what we eat, but how the food is eaten as well. Similarly, as we proceed through the events of each day, we construct narrative to make sense of our lives; we conceive the script and enact the roles chosen. The events we choose to include in our narration become the narrative of ourselves—become ourselves. Whatever else autobiography might be, it is a narrative of the moment. The work is, of course, my construction and I ought to take responsibility for it. My autobiography is *my* story but not *the* story. It is inconceivable to me that Henry Adams could write his entire autobiography and not recall in it that his wife had committed suicide. But I can acknowledge that that event was not part of the story Adams meant to narrate *at that time*. I am, however, free to interpret Adams's autobiography informed by this remarkable silence. In that narrative, I can understand that he inhabits that silence; I construct my story of Henry Adams from the story Henry Adams narrates for me.

Perhaps we all produce several editions of our autobiographies suitable for various audiences. But what all these various narratives have in common is their identity as autobiography: stories of self. I think that what is true of autobiography for people, is true as well for nations. Yosef Yerushalmi, a Jewish historian whose work *Zakhor* explores the relationship between the writing of Jewish history and the experiencing of Jewish memory (1982:99) says, "The choice for Jews as for non-Jews is not whether or not to have a past, but rather—what kind of past shall one have." We make our own histories, Marx cautions, but not on conditions of our own choosing. Stories are us.

But I have also been Marxist enough to know that alienation occurs when we cannot tell the story we know, or when the story we mean to tell is silenced or denied, or when we cannot know or are denied our participation in the story. So much talk of education and curriculum has been connected to ideas of resistance. Resistance is one response to the silencing of our story. The title of Herbert Kohl's book (1994), *"I won't learn from you," and other thoughts on creative maladjustment,* gives some ready definition of the effects resistance might produce in the classroom in which the students experience alienation. Perhaps the engagement in study is only possible—is only effective—when it makes possible a connection to the story. Perhaps productive study is the immersion of the student in the story. When we learn, we become a participant in the story told, and the story then becomes available to our own life narratives. The story that we may join enlarges our life by multiplying astronomically the participants in it and increases the threads of potential development for our life even if our material condition denies access to them. I tell my students that if nothing else, curriculum is the story we tell our children. What we teach our children is the picture of what we mean to tell our children the world is like. Today Emma came home and told me that the substitute teacher made her class walk down the halls in absolutely straight lines. What was that teacher telling our children about the world? In story, our lives may become comprehensible. Today, my daughter came home from school confused. And so we constructed narrative to make sense of her experience. Emma was not any more pleased with the story's explanation, but she was at least satisfied with it.

✡

Here is the next story from the Gemara presently under discussion. It departs from the metaphysical consideration of existence raised by Rav; neither does it address matters of social justice and its execution. Rather, the Rabbis move dramatically and almost comically to the topic of sensuality and sexuality. The story itself makes sense, but not certainly in its context.

> When Rabbi Yishmael the son of Rabbi Yose and Rabbi Elazar the son of Rabbi Shimon happened to meet each other, a pair of oxen could pass between them and not touch them. A certain [Roman] matron said to them: "Your children are not yours!" They said to her, "Theirs is greater

than ours." "All the more so!" There are some who say that they said to her thus, "For as the man is, so is his strength." There are [others] who say [that] they said to her thus, "Love presses the flesh."

But why did they respond to her? But surely it is written, "Do not answer a fool according to his folly"! So as not to spread evil talk about their children.

Rabbi Yohanan said, "The limb of Rabbi Yishmael the son of Rabbi Yose was like a water-skin measuring nine kavs." Rav Pappa said: "Rabbi Yohanan's limb was like a water-skin measuring five kavs. And some say: measuring three kavs. That of Rav Pappa himself was like the baskets of Harpanya."

Talmud may be holy text, but I am certain that this story would not be selected for William Bennett's *The Book of Moral Virtues*. This story follows immediately Elijah's command to Rabbi Yishmael, who was arresting Jews for the King. You will remember that when Yishmael asked Elijah "What can I do," Elijah told him he might flee to Laodicea. This exchange followed the longer story of the scholar Elazar, whose employment as government mole, as it were, led to a series of issues concerning ethics. But with the opening words of this present story concerning overweight rabbis, I find myself immersed in salacious talk about the dubious legitimacy of the children of these very same Rabbis who have been engaged in issues of life and death and life after death. From the present talk it would seem that the Rabbis' girths are too great to permit them to engage in sexual intercourse. The Roman matron doubts the legitimacy of their children. The Rabbis respond that their wives' bellies are even bigger than their own and therefore, intercourse is, indeed, possible. Upon hearing this, the Roman matron is even more convinced of the impossibility of sexual relations. On the other hand, the men boast, the abundance of a man's flesh is symptomatic of his capacity for sex. Finally, the Rabbi's acknowledge that the love they feel for their wives is able to overcome any physical impediment to intercourse.

I wonder where the talk about the wicked and the righteous has disappeared to, or what has happened to the issue of labor relations. The characters have not changed but they seem to be in a different play. But as Talmud is story, so I must construct the plot.

I attended an Individual Education Plans (IEP) meeting today for Thomas Herschel. Tommy is a sixteen-year-old child in ninth grade at Memorial High School in a major mid-Western city. Memorial is a big school, serving a racially mixed population of almost two thousand students. The school also houses the Fresh Start program, which places youthful criminal perpetrators in the schools. Thomas was once relieved of thirty dollars by some very mean-looking classmates. Everywhere I looked, there were guards and security personnel. Regularly, I would hear over the public address system or on a two-way radio: "This student is walking the halls unattended;" or, "Ms. Williams reports that a certain student left class to go to the bathroom ten minutes ago and has not yet returned." I could not distinguish this school from any prison I have imagined.

Individual Education Plans are legally required for each student classified with special needs. Thomas is presently classified as Learning Disabled. On the typical standardized IQ tests, he scores between low average and average level; by intellectual measures he should be capable of working at a level appropriate to his age and grade. However, Thomas functions several degrees below grade level (there are, I suppose, several degrees of separation between him and the average student) and is eligible for special services according to federal law PL 94-142. I am glad that Thomas is eligible for special services. I think everyone should be. When Emma was in first grade, she had a remarkable sense of story and a wonderful ability to grasp a storyline and intent, but she seemed incapable of attaching herself to print. "What is the first sound you hear in c-a-t?" we might ask. And she would calmly, even proudly, answer "m." She lacked, in academic parlance, phonemic awareness. She was placed in a Chapter I reading group available at her school as a result of the large number of disadvantaged children in attendance. When first I heard she was assigned to Chapter I reading, I reacted as if I had been told she was afflicted with terminal cancer. When finally I calmed down (and fortunately I engaged in this episode in isolation) and considered Emma's situation, I saw her sitting in a small group with a caring, dedicated and highly competent reading specialist who was herself working in concert with a caring, dedicated and highly competent first grade teacher. I wished everyone were assigned to Chapter I reading groups.

I have never been in a class in which everyone was working at this magical grade level: it is certainly not the case at the university that everyone's functioning is at a uniform level. Certainly in the elementary and secondary schools there is inevitably a wide range of working levels that exist within any single class of the students. But, a classification of learning disabled is made when a student performs a minimum of two grade levels below her present position, and there is no other physical condition that might explain that deficiency. That is, the classification refers to the student's learning condition and not to the student's physical state. When we describe someone as learning disabled we ascribe the deficiency to the student, yet another way we in the twentieth century have learned to blame the victim. Thomas is not living up to someone's idea of his potential. His doctor prescribed for him Ritalin. When he does not take it, his teacher says, he is "off the wall."

It is not that we should not have expectations for our children. How could we operate in the course of a day without an expectation of tomorrow? I cannot see my children outside of their existence in the future. To all of our detriments, I too often force them into my vision of their future. For example, this morning I fought with my daughter about wearing clothes appropriate to the Wisconsin weather and about brushing her hair before she left the house for school. There is so much I/we want for our children that it is possible to hover over them every single moment ensuring that they move in the path we desire for them. I am overly conscientious—perhaps neurotically so—about her reading and her homework and her violin practice. I too often cross the line of imposing my desires on her and denying her the freedom to discover her own. It is, I think, a daily struggle, and this morning was a particularly unpleasant one. It does not derive from a lack of love or caring, but perhaps from its opposite. If I loved wiser, I would love not less well, but less strictly.

But I resist assigning Emma or Anna Rose to grade levels and then operating as if they had to stick to them or treat the children as if these levels were real. These levels, like floors in a department store, are an architecture that maintains separation and distance. Fourth grade: multiplication tables, Wisconsin history, electricity, lingerie. I learned from Lev Vygotsky (1988) that education not only precedes development but actually causes it. What do grade levels matter if growth is the end and there is no final resting place, so to speak. Study is, after all, its own reward. If we had paid attention to someone's notion of reading levels, Emma would have missed Mildred Taylor and Jane Yolen and Anne Frank and . . . well, so much. I am not certain that my children could ever be that consistent. Nor perhaps would I desire that they be so. While Emma reads books far more advanced than her elementary school placement, she remains committed to her fingers for certain mathematical operations. On the pre-kindergarten screening tests, Anna Rose was assigned a 67. Sixty-five is the minimum grade for admittance to kindergarten. I asked the tester: "Did you ask her to write her name or the names of any family members?" "No," she said, "that isn't part of the test." Anna Rose can write not only all of our names but the names of her friends and day care providers as well. She can write almost the whole alphabet. "Did you ask her to count as high as she could?" "No," she said, "that isn't part of the test." Anna Rose can count almost flawlessly to a hundred. "Well, did you ask her to read from a book?" "No," she said, "that isn't part of the test." Anna Rose can read the Maisy books all by herself. Ask her, she'll tell you.

I think Anna Rose performed below grade level on that kindergarten screening test.

Thomas actually seems to me to be borderline autistic. Earlier, he was indeed tested for autism, and though he met several of the criteria for the classification, he did not give enough quantitative evidence for such definite classification. Nonetheless, he avoids eye contact; his expressive language is seriously deficient; he makes little or no effort to engage in social activity. He does not venture much out into the world; though he loves the movies, he does not know how to get to one on public transportation. I did not see Thomas initiate any conversation, not even with his overwhelmed mother. She does his talking for him. When a question is addressed to Thomas, the addressor usually answers for the addressee. "Don't you want to go to the movies, Thomas?" "Of course you would, Thomas, you love the movies." "Would you like to learn to ride the bus, Thomas? Sure you would!" "Thomas, don't you have anything you want to say." (This is actually a common problem. Talk to any young child in the presence of their parents or grandparents and observe how often the adult speaks for the child!) The school reports that Thomas talks with almost no one, and then only in brief and whispered half-sentences. When I try to engage him in talk, he stares at me as if waiting for me to say something else, or else, as I have said, someone around the table quickly answers for him. He tells me that at lunch he sits alone: in such a big school, it is hard to imagine how he finds a table to himself. His teacher says that he has no communication with anyone in class.

Many of us have such inclinations. We sit in the tavern apart from the crowd and stare ahead. Or read our newspaper. We fall asleep. We avoid contact. But Thomas is supported in his taciturnity by the structures of the school. Despite national efforts for inclusion and despite the mandates of the Americans with Disabilities Act, for almost the whole day Thomas is segregated from the student body in a room specifically assigned to the learning-disabled students. His teacher says that Thomas doesn't ask many questions: when he needs to know the assignment he just goes up to the front of the room and looks in the teacher's plan book. "I leave it open and students independently go in there regularly to check on what they should be doing. Thomas even writes the assignment on the board." I suggested that if Thomas had to ask another student for the assignment, he might learn the benefit of communication. He might learn how to communicate. When the meeting was over and the bell rang, other classified students entered the room and one sat down at the computer and began playing a game. "Hey, Thomas," want to help me play this game." "No, "I'm busy." When Thomas declined to join his classmate at play, it was because he was writing the assignment on the board for the teacher.

Thomas produces comic books. He draws characters in vibrant colors and gives them assorted names. They are all based in science fiction. Every character is veritably the same: posed in identical postures, each stands flat against the page, limbs splayed out stiffly with neither muscle tone nor articulation. The bodies are all geometrically shaped and the heads small and anonymous. As with the arms, the legs are straight and stiffly drawn. There are no faces. Only masks. Every one of the characters is Thomas. There is absolutely no text created; only names attributed to the characters. He has already produced three completed books.

During the IEP meeting, it became apparent to me that no one previously knew about Thomas's work; his mother arrived carrying several three-ring binders and a sketch pad (in which Thomas was busily engaged on a work in progress). Each person at the table—his classroom teacher, a transitional specialist, and a transportation specialist (I thought there should have been more—at least the school psychologist!) was asked to examine Thomas's work. "Look at Thomas's work," his mother said, as she hopefully and proudly pushed the volumes across the table. Thomas sat abashedly at the meeting. "These are great, Thomas." "Wow, Thomas, this must take a great deal of time." Thomas's mother says that he works on these comic books during much of his free time. This drawing seems to be his one passion.

Thomas is constructing himself when he draws these figures. I thought to myself that this drawing was the only way Thomas assures himself that he exists. No one else ever sees him. He passes between them without being recognized. At the meeting, the transitional specialist mentioned that Thomas should be placed in a work environment next year where he might be involved in, say, preparing school lunches. There were other activities mentioned in which he might engage, but I have now forgotten them. I asked if there weren't a site in the school district where Thomas could be engaged in graphic design and drawing given his focus on art and drawing. "Well, yes, there is such a site but it isn't necessarily structured as a "1930" site and Thomas might be better served with structure right now." I didn't know why—it seemed to me that Thomas is presently suffocating in structure. But no, that metaphor is inappropriate; Thomas is completely lost in the structure right now. No one can see him. He passes right between the overweight rabbis and under their extended and self-satisfied bellies. The lions ate Thomas.

✡

Perhaps there is a relation between the passion for sex and the passion for study. Martin Buber (1947) tells this Hasidic tale:

> A learned man once said to the rabbi of Rozdol: "It seems to me that the condition of being a zaddik (a wise person) is the greatest of all lusts."
>
> "That's how it is," the rabbi replied, "but to attain to it, you must first get the better of the lesser lusts."

That is, I need not turn to asceticism to be a scholar, but must master the other appetites so as to leave space for study. Perhaps that is why Elazar considers those asleep in the tavern as thieves if they be neither workers nor students: slave to the lesser lusts, they steal from themselves the plentifulness of the narrative.

And there are some who say the Rabbis weren't talking about skin or limbs at all—they were talking about the size of penises.

✡

> Rabbi Yohanan said, "I remain from the beautiful ones of Jerusalem." He who wishes to see Rabbi Yohanan's beauty should bring a silver goblet from the refiner's shop, and fill it with the seeds of a red pomegranate, and place a wreath of red roses around its rim, and set it between the sun and the shade.
>
> That glow is similar to the beauty of Rabbi Yohanan.
>
> Is this so? But surely the master said: Rav Kahana's beauty was similar to the beauty of Rabbi Abbahu; Rabbi Abbahu's beauty was similar to the beauty of Jacob our father; the beauty of Jacob our father was similar to the beauty of Adam. But he did not include Rabbi Yohanan!
>
> Rabbi Yohanan is different because he did not have a beard.

I think that when I was in the twelfth grade I fell in love with Mr. Raymond Matienzo, my English teacher. He was probably young then, perhaps even fresh out of university, though from my perspective as a high school student, he seemed quite wizened and experienced. And though he did have a beard and was, therefore, perhaps not as beautiful as Rabbi Yohanan, I was thoroughly enamored with him. If there was a sexual component to this love, I was then oblivious to it; now, I can only tentatively ascribe some homoerotic passion for the man who taught me the depths to which literature might be plumbed. I had always been a reader, having discovered in books the legitimacy of emotion that my life seemed to deny. In books, I confirmed the passion I experienced but could share with too few in the world, and certainly not with any one in my immediate family. We were guarded and loathe to display emotion. I remember Mr. Matienzo, however, being passionate about literature; he had a high voice that sang even in its lower registers. When he read from books, his face glowed, and he gesticulated wildly during dramatic readings. For Mr. Matienzo, doing literature was about living life and not about writing book reports. It was he who taught me how to experience life in literature, and it was Mr. Matienzo who made my readings welcome; it was he who sanctioned my experience in reading. In loving him, perhaps, I learned to like myself.

Perhaps not unconnectedly, it was Mr. Matienzo who introduced me to existentialism. He had taken his senior English class to see a university production of Samuel Beckett's *Waiting for Godot*. I do not know that I understood very much of the play, nor will I say I was not at times puzzled and bored by it, but I believe I was intrigued by the presence of those two comic characters alone on the bare stage waiting for someone who never arrived. Perhaps I read into *Godot* a mirror of my own adolescent angst. For my senior project for Mr. Matienzo's class, I chose to study what was then being referred to as The Theater of the Absurd. I read the plays of Edward Albee and Eugene Ionesco and Jean-Paul Sartre. I read and did not understand Camus' *The Stranger*; I studied assiduously Martin Esslin's book, *The Theater of the Absurd*. I think that this volume is the first in my library in which underlining appears. In 1965, as part of the experience of this project, I took my father down to the Cherry Lane Theater in Greenwich Village to see performed in a theatrical double feature productions of Edward Albee's *The Zoo Story* and LeRoi Jones's *The Dutchman*. My father was appalled by the seeming lack of sense in Albee's drama and with its final violence, as he was abashed and shocked by the vituperative language in Jones's play. Later, my father told me he saw a rat run across the rear of the stage during the first play, and following that sighting he had kept his feet raised off the ground for the remainder of the afternoon. I don't recall our discussion of the plays.

I was enthralled with the possibilities I imagined in existentialism. The emptiness of the absurd universe nourished my romantic teen-age melancholia and adolescent rebellion; the black turtlenecks I began wearing were emblematic of my soul, darkened not by the presence of evil but by the absence of light. I began frequenting the coffee houses of Greenwich Village where folk singers and folk poets articulated my inchoate rage, confirmed the meaningless of the universe and bewailed the injustices of the world I had had no part in forming. I learned the language of codes and of subterfuges. I learned—learned again, for in Hebrew school I had first been instructed that no word need be taken at face value—that everything could and did mean something else. Even poor "Puff, the Magic Dragon" who lived by the sea (!) with his friend Jackie Paper (!) was condemned to a life of cocaine distribution and drug abuse. Existentialism confirmed the victimhood that my adolescence and world conditions had imposed. I filled the depths of the abyss with adolescent despair, rebellious anger and incessant reading. I sang the virtues of *The Bald Soprano*. Against the monotheism of my father's Judaism, I erected the blankness of Raymond Matienzo's existentialism and the absent divinity of Ionesco and Beckett. To my father's world implicit with meaning, I opposed a world without sense and order. In existentialism, I grounded my revolt. "Nothing to be done?" Of course, there was something I could do: I could embrace the absurd and call it my own. Raymond Matienzo was my Elisha ben Ayuda, the apostate, the Other. Raymond Matienzo was my Don Quixote. Raymond Matienzo was my Beelzebub. He was beautiful and I loved him.

How can we not love our teachers? When we look at them, we see creation. Isn't that what the master meant when he compared the beauty of Rabbi Kahana and Rabbi Abbahu to that of Jacob and Adam. It is through this lineage that the Jews attribute their ancestry; it defines their election. But I wonder now, how can we know the beauty of either Adam or Jacob? There are no photographs nor framed portraits! Perhaps the better question seems to be: of what does their beauty consist? Maimonides tells us that when the Torah says that Adam and Eve were made in the image of God, it refers not to a corporeal presence, but to their capacity for intellectual apprehension. Adam and Eve were given an overflowing and I suppose pure, intellect. It is this quality that Maimonides attributes the identification of Adam and Eve being made in the image of God. It is from their capacity for intellectual apprehension to which their beauty may be ascribed. It is through the intellect, says Maimonides, that the false and true may be distinguished. Adam and Eve originally possessed this power of discernment. As an adolescent in Mr. Matienzo's class this distinction was central—and certain—to my life.

I am glad that I am no longer either that wise or that old. I see Mr. Matienzo's wry smile even now, and I know it shone ironically on my innocence. I think he long before had abandoned his faith in the knowledge of the true and the false.

Maimonides (1963) tells us that for the disobedience in the Garden, Adam and Eve were punished, as would be all human beings, by being deprived of the capacity to contemplate only the most spiritual things—that which partook of the true and the false—and were denied the ability to distinguish between them. Rather, following their taste of the fruit of the Tree of Knowledge, Adam and Eve learned to consider, instead, the general things of the world, the diurnal objects that comprise our existence, and they were constrained to consider these things not with regard to their truth or falseness, but with the awareness of only their goodness or badness. The latter judgments, however, were matters of mere opinion and inclined Adam and Eve towards satisfaction of their desire and their imagination. Maimonides considers that by their consumption of the fruit of the tree, Adam and Eve had striven to satisfy their physical and sensual appetites; for this desire the couple were chastened by being deprived of their purely intellectual apprehension, now tainted by an awareness of good and evil. Maimonides says that when Adam and Eve "disobeyed and inclined toward [their] desires of the imagination and the pleasures of this corporeal senses—inasmuch as it is said: *that the tree was good for food and that it was a delight to the eyes*—[they were] punished by being deprived of that intellectual apprehension." No longer were they able to consider the world in abstract and ultimate categories of the true and the false, rather, they now were obliged to engage with the world in all its mundane complexities. Notions of the false and the true give way to decisions of good and evil when the necessities of life must be labored for by human hands. The absolute certainty distinguishing the false from the true had to give way to the muddied materiality in which is mixed the good and the bad and where each is tainted with the substances of the other. The cold certainty of intellectual apprehension gave way to the warm beauty of human struggle and uncertainty. I think the beauty of Adam and Jacob and of our teachers derives from their own engagement in human struggle and doubt. I never doubted that Mr. Matienzo had eaten of the fruit of the Tree of Knowledge. I mean, he had read *Waiting for Godot!*

I do not know, indeed, if the disobedience of Adam and Eve wasn't, in fact, an intentional provocation: the value of our lives comes, after all, from its brevity and its struggle. Perhaps Adam's beauty, as that of Eve as well, consists in their willingness to engage in life. Their expulsion from the garden appears less to embitter than to energize them. Adam and Eve utter no recriminations or voice regret following their expulsion from the garden: indeed, I speculate that their beauty derives from the full acceptance and embrace of their lives. They do continue to be fruitful and multiply. Perhaps humans were expelled from the garden so that they could embark on the life we know; Adam and Eve were expelled from Eden to prevent their eating of the Tree of Life and to live forever, thereby losing forever the wondrous obligation of involvement.

The expulsion from the Garden interpreted in this light is not punishment, but salvation. Our life outside the garden is what makes us human: it is the engagement with effort. To be human is to struggle with the earth—struggle derives from our mortality. Our mortality results from our desires and imagination. Maimonides says that the desire to eat of the fruit gave to human beings the knowledge of good and evil, the result of imagination and opinions, and took from the human the sole pursuit of the false and the true, measures of the absolute. It is why, Maimonides says, the Torah says that after Eve and Adam ate of the tree "the eyes of both were opened, and they knew that they were naked." The very way Adam and Eve thought had changed as a result of their deed though the world remained the same: now, however, their nakedness was no longer acceptable to them. Indeed, our opinions change with our engagement with the earth—what one may at first consider good may later be defined as evil, but then, the former always contains the latter and vice versa. Perhaps it is the admixture of good and evil that makes us human, and therefore beautiful. It is, after all, impossible to work the earth and remain clean.

Adin Steinsaltz (1992) says that the sense of being human begins from where the human himself issued: the dust of the earth. That is why the Torah says that Adam and Eve were banished from the garden that they may "work the soil from which he was taken." That engagement with the earth is an engagement with the very substance of humanity: human beings must be of the world. In Beckett's *Waiting for Godot*, however, Vladimir and Estragon do not move; they wait. They detest their struggle with the earth though it is inevitable that they do so. Unlike Adam and Eve, Vladimir and Estragon hate their life. Mr. Matienzo, my teacher, taught the glory of the struggle; he does not appear unstained in my thoughts as does, say, Mr. Martin, my twelfth-grade physics teacher, whose clean white shirt remains in my mind forever spotless. When Mr. Matienzo cried, "I fall upon the thorns of life, I bleed," I saw the red stain sullying his shirt and believed he reveled in the trial. The struggle was his achievement.

Abbahu's beauty is said to be also like that of Jacob. Jacob, too, works the soil—he binds himself uncomplainingly for twenty years to his Uncle Laban despite his uncle's deception and exploitative practices. Jacob, too, like Adam and Eve, must leave the security and safety of his family's garden, as it were, where all was certain and secure, and must venture into the world to labor with the earth. Jacob's dream of the angels ascending and descending the rungs of the ladder suggests to him that God lives, indeed, in the world: upon awakening Jacob allowed that "God was in this place, and I, I did not know it." I think Jacob learns human humility by his struggle with the earth. Jacob learns value by working the soil. It makes him beautiful. Working the soil means struggling and sorting and plowing and harvesting. None of these activities can be accomplished without intimate contact with the soil. All that Jacob finally gains in his life derives from his work. It is not the true and the false with which Jacob deals. Jacob often errs and is subject to deception: he experiences the good and the bad, and he must mourn *and* rejoice. In his humanity, which derives finally from his engagement with the earth, Jacob is beautifully human.

The teacher Abbahu's beauty derives from that which attaches to Adam and Jacob who gained their humanity from their engagement with labor. I think Mr. Matienzo gave me the earth, which I had then to work. My beautiful Raymond Matienzo taught me the efficacy and the dignity of working in the fields. He taught me the beauty of that effort.

But, sometimes, I remember that I am on sabbatical and I am forbidden these activities.

Ideas, I think, always come to us embodied; perhaps we cannot love the idea unless we also love the body that makes its introduction. Then, perhaps, the body becomes beautiful. If Raymond Matienzo offered me what I innocently thought were visions of truth and falsehood, he also inspired in me knowledge of good and evil. I have long since abandoned all hope of discovering the former, but I have never ceased to learn to distinguish between the latter. And it is, I believe, this quest that attaches me to my children and to students.

Perhaps it is to Adam and Eve to which the origins of scholarship might be ascribed; it is, of course, their act that led to the distinction between good and evil. Since I cannot discover what is true or false in the world because of my engagement with the corporeal and the diurnal, then perhaps I can seek in my acts what is good in the world and to abjure what is evil. I might in the home and in the classroom begin, thus, the repair of the world—*tikkun olam*. It surprises me to consider that I may owe my life and my passion to Adam and Eve's disobedience. How can I condemn them? How can I not love them and think them beautiful. I think Raymond Matienzo's beauty derives, indeed, from that of Adam and Jacob.

But the classroom is inhabited not only by students who love the teacher; teachers, too, must experience a passion for their students. I know we often say that we love our students, but this is an idealistic and altruistic love in which I do not believe. If in their faces we see merely reflections of ourselves, then we do not see the students at all and it is ourselves we love. Or if we see in the students only empty vessels that we desire to fill, then it is our knowledge we love and not our students. Rather, I think we must love our students as the Levinasian Other—the one we cannot know and whose presence ensures the existence of the world. But if we love them in this manner, then how could we ever ascribe to them a failing grade? If we do not love all of them, then which of them is it that we do love? Perhaps I love only the ones who most fulfill me. This is again a selfish love. Another story:

> One midnight when Rabbi Moshe Leib was absorbed in the mystic teaching, he heard a knock at his window. A drunken peasant stood outside and asked to be let in and given a bed for the night. For a moment the zakkik's heart was full of anger and he said to himself: "How can a drunk have the insolence to ask to be let in, and what business has he in this house?" But then he said silently in his heart: "And what business has he in God's world? But if God gets along with him, can I reject him?" He opened the door at once and prepared a bed (Buber, 1947).

When I open the door to the classroom, who am I to judge the character and legitimacy of she who would enter? If God accepts her, then how can I reject her? If she would enter, I will welcome her; perhaps I must even love her. My self-absorption in my mystical teaching is well interrupted by my concern for the Other.

As for the bed I would prepare—there are a legion of rules and regulations forbidding that! Thanks goodness. I think it is a complex conversation to say we love our students.

And what do we know about the students who love us? I often hear teachers say, "I don't care if they like me I just want them to learn from me." Do we seek to be loved when we teach, or does love seek us in the thrall of study? Perhaps without love there is no possibility of study and learning. If I do not inspire love, then how could anyone learn from me? I wonder if perhaps one must be loved in order to teach? I wonder if perhaps one must first love in order to learn.

Perhaps the connection between love and study teaches why we have these ribald, even bawdy and somewhat obscene stories about the Rabbis. They are flesh and spirit, and when we attempt to separate the two, as we do too often in schools, we deny our own humanity and that of our students.

✡

Rabbi Yohanan would go and sit at the gates of the ritual bath. He said: "When the daughters of Israel come up from [their] obligatory immersion, let them meet me so that they may have beautiful sons like me, students of the Torah like me." The Rabbis said to him: Are you not afraid, sir of the Evil Eye?" He said to them, "I come from the descendants of Joseph, over whom the Evil Eye has no power, as it is written: 'Joseph is a fruitful bough, a fruitful bough by a well.'" And Rabbi Abahu said: Do not read "by a well," but [rather] "above the power of the eye." Rabbi Yose bar Hanina said: From here: "And let them multiply abundantly like fish in the midst of the earth."

I wonder if there is not a significant element of voyeurism implicit in teaching. I do not refer to the criminal variety or even to the socially unacceptable form of voyeurism. Rather, I refer to the commonly held idea that the best practices of teaching amount to sustained and informed policies of "kid watching," observing students assiduously, carefully, continuously, and yet remaining, as it were, unseen. I do not know that teaching could be accomplished without such canny and intent scrutiny that must all be carried out with not a little secrecy. It is prerequisite that we assiduously and continuously monitor and interpret the behavior of the children in our class. In academic circles that process of research is today referred to as "authentic assessment"—the mindset of teachers who are constantly observing students to assess what they already know and where they need next to learn. The activity of observation is implicit in the language we employ to describe our behaviors: we *watch* intently the actions and reactions of students in constant assessment of their progresses so that we might make plans for their continued development; we bemoan those who, through our own *lack of vision* and our negligence, have fallen through the cracks. We berate ourselves for *losing sight* of our primary functions: to see the child in the student. Mustn't there always be a bit of the voyeur in my gaze? Isn't there always in me a desire to look unseen, to steal furtive glances at my active charges busy at study and innocent of my evaluative gaze, which assesses each behavior and judges its conformity to expectations and stated objectives? Don't I continuously survey my students for evidence of newly planted seeds of myself in them? Or more dangerously, don't I observe my students in their tentativeness and insecurities, in the vulnerability they inevitably acknowledge by the (un)willing exposure of their lack of knowing. Indeed, in all of their nakedness?

I think that we teachers often walk into the classroom as socially sanctioned voyeurs. Students come to us, as it were, naked. By the very structures of the educational establishment, students are commanded to stand before us veritably unclothed, helpless, acknowledging their deficiencies. Not from the bite of the fruit of the Tree of Knowledge, but from their attendance in school do students become ashamed of their nakedness. In our classrooms, they are exposed before our enlightened gazes, frightened of our power, intimidated by our learning, and shamed by the revelations of their vulnerabilities. That is, in the unequal power relations that exist in school, teachers by their gaze effectively pin students to their seats. We ourselves stand behind our desks hidden from full view, hidden, as it were, by our fig-leaf aprons, and peer at the discomfiture we provoke. We cloak ourselves in our texts and tests and worksheets and standards, and we watch assiduously or blankly or meanly or ignorantly at the revelations our students proffer in their studies. We accept haughtily their praises and their tentative offerings, and we stare from a safe distance at their submissions. We insist that students perform before our gaze for their own good. Socratic teachers ask questions knowing in advance the path they have chosen and upon which students should race, as they gaze uninterruptedly on their efforts. We order our questions according to certain taxonomies and often cannot observe an alternative response. Or we ask questions for which there are no clear answers, and students, trained to respond but terrified of the vulnerability in their exposure, wriggle in their seats before our gaze. Teachers lecture while students sit busily occupied taking notes: we monitor those who write responses to our questions and eye those who in our classrooms do not seem to us acceptable. We pick up their notes and read their private communications. We assign reading and then evaluate the quality of the selves that have been produced in that reading based in our own personal preferential understandings. We examine the exposed souls in the writings we collect and graffiti them with red ink and with obscene grades. We interview and interrogate. We ask questions for which they have forgotten—or perhaps never anticipated—an answer. We fix them to their places denying movement or companionship. We test every aspect of their lives and quantify their complexities in stanines and numbers and grade levels and percentages. We humiliate by this exposure even as we seek to facilitate. We offer the promise of clothing to naked souls. We do all of this sanctioned by everything we know to be education; those who gaze the best are best rewarded—they are named "teachers of the year."

Perhaps when we look out at the classroom we recognize the nakedness of our students and we are, ourselves, ashamed at our voyeurism. Perhaps we recall our own shame in the classroom. I remember, as it were, my own nakedness by uncovering theirs. Could that be a source of my discomfort in the classroom?

Perhaps the classroom is not unlike the mikvah before which sits Rabbi Yohanan. In Jewish custom, the mikvah is a ritual bath in which men and women purify themselves. Isn't that what education is meant to accomplish? We are constantly adjured to make ourselves a better person, a better community and a better nation. We are adjured to perfect ourselves immersed in the environments and culture of the school. When we leave the school, we have been hallowed; our misunderstandings have been cleansed away and we have been washed in the clarifying waters of reason. In the hours preceding Shabbat, the impurities of the workweek are washed away in the mikvah, even as our minds are supposedly cleansed of false opinions in schools. After menstruation and before intercourse, women immerse themselves in the ritual waters of the mikvah in preparation for lovemaking; men go to the mikvah after nocturnal emissions. It would be haughty to stare at those entering the mikvah even as it would be a defilement to observe unobserved the purity worn by those exiting the mikvah. The Rabbis are rightfully alarmed at Rabbi Yohanan's effrontery.

Schools, too, are defined as transitional mechanisms where one is made ready for the "real world." In the mikvah, the body is immersed so that the soul might aspire. Bodies are what we try to ignore in schools, though we attempt to control them; it is a sin to gaze at the bodies as they emerge naked from the ritual bath, but we stare assiduously at the students who frequent our classrooms. We assume our privilege and ignore our potential to shame. Or we translate this power into a virtue. And so the Rabbis are surprised that Yohanan is not afraid to expose the nakedness of the daughters of Israel. Isn't Yohanan afraid of the Evil Eye? I am myself terrified of students: I stand before a class and am unnerved by their beating hearts. On the one hand I know that the students look to me as the one who knows. They are terrified of my power. I am terrified of my power. I am frightened that all of my knowledge is inadequate to the task that abides in the classroom. I am terrified that I will not know what to teach these students. I hide my fear behind my gaze. I stare at them and make them naked. I watch them emerge, as it were, from the mikvah. Shouldn't I be fearful of the Evil Eye?

And so it interests me to hear Johanan explain that because he is descended from the line of Joseph, he is immune to the powers of the evil eye. Joseph is a complex figure in the Torah, and it is not my intent to deal completely with him here. But certainly, Joseph's position as teacher and the complexities that this position encompasses appear relevant to my present musings. For Joseph's mission requires that in his capacity as teacher he find evidence of learning even as he gives evidence of having learnt; that he observe out of love and yet in his voyeurism chasten those

he loves; that he be loved and yet inflict pain upon those who express
their love to him. Perhaps it is his acknowledgment of his learning that
makes Joseph so humble in his role as teacher; perhaps it is this humility
that authorizes his voyeurism. Perhaps it is Joseph's willingness to uncover
himself upon which rests his power to educate. Perhaps it is this final
humility of the teacher-learner that permits the learner-teacher to sit outside the gates of the mikvah and be not subject to the Evil Eye—to the
hatred of the students.

In their jealous rage over their father's unequal affections and their
legitimate anger over his arrogance, Joseph's brothers attempted to do
away with him, but at the earnest pleadings of their oldest brother,
Simeon, they contented themselves with merely selling Joseph into slavery; Joseph survives his ordeal. Separated from the family dynamic, no
longer believing in his final ascendancy over them that he had foretold in
his dreams and removed from the privileged and self-centered space that
the family, primarily that his father, Jacob, had offered him, Joseph must
acquire and utilize skills from a position free of almost all self-interest. He
must now interpret— without ulterior motive—the dreams of others,
including those of Pharaoh. Joseph's limited insights gain him reputation
and power. Having predicted economic trends, Joseph is made viceroy
and given extensive powers to control agricultural and economic matters
in Egypt to prevent catastrophe.

Famine brings Joseph's brothers to his court. They have descended
into Egypt for food to relieve the consequences of starvation throughout
the land. When they come to Pharaoh's court, they do not recognize in
the viceroy their brother Joseph, but Joseph "recognized his brothers,"
and from his privileged position—privileged in knowledge and power—
Joseph tests his brothers to see what they have learned—how they might
have changed over the course of their histories.

In exchange for food, Joseph orders the brothers to leave behind one
of their own and to return again to Egypt with their youngest brother,
Benjamin. Joseph offers his brothers an opportunity to save themselves at
the expense of others. And his choice of methods is brilliant: it recalls to
the brothers their earlier cruelty to Joseph: "Indeed, we are guilty concerning our brother inasmuch as we saw his heartfelt anguish when he pleaded
with us and we paid no heed; that is why this anguish has come upon us."
Even as the brothers discuss their chastened and guilty state, Joseph
unknown to them, listens to their conversation. And Joseph turns away
from them—he has been the voyeur; he has overheard, and he suffers with
their pain. Nonetheless, the test must continue: "He turned away from
them and wept; he returned to them and spoke to them." And Joseph

takes his hostage; as had been Joseph, so too, is Simeon left behind. In this test, Joseph would know how his brothers have changed. Joseph is a dreamer. Joseph has been his father's obvious favorite—the infamous coat reflective of the special love Jacob bestows on him—and the brothers respond with an understandably bitter jealousy to the father's one-sided affections. Perhaps as a result of his father's affections, Joseph had become haughty and proud; in his arrogance he not only dreams, but is prideful enough to tell his dreams. And Joseph believes not only the literalness of his dreams, but his interpretations of them as well. He assumes that what he knows is all that there is to know. Joseph lacks all sense of humility; his arrogance, which may stem from his father's overly effusive attentions, makes Joseph insensitive to the feelings of his brothers. Joseph lacks a sense of tact. When he recounts to his brothers the dreams that portend his ultimate ascendancy over them—and their subjugation to him—they hated him more for his talk than his dreams. Joseph never denies that his dreams portend his dominance over his brothers: he belittles his brother with his dreams. It is Joseph's role to be a dreamer, but he must first learn to keep his dreams to himself. Joseph must become a learner.

We all dream. We are all dreamers. Don't I always cherish dreams for my children? As a teacher, I stand in the classroom and I dream. But Joseph tells his dreams as if he knew also the means by which the dreams would come true. Now, Joseph's dreams prove true, but in a way Joseph could never have foreseen, nor in a manner for which he might have planned. Joseph comes to understand that he has been directed by a destiny that exceeds his capacity to control and that he has been subject to a power he does not comprehend. He has served purposes other than his own. In revealing himself to his brothers he offers them comfort: "And now, be not distressed, nor reproach yourselves for having sold me here, for it was to be a provider that God sent me ahead of you . . . Thus God has sent me ahead of you to insure your survival in the land and to sustain you for a momentous deliverance." It is wise to hold our interpretations in abeyance that we not close off possibilities for our futures. We do not know the ends of our dreams nor the ultimate purposes we serve. As teachers we would do well to recall the story of Joseph; it would remind us of our position as learners. It would keep us humble. We teachers would be descended from the tribes of Joseph and be not subject to the Evil Eye as we sit gazing outside the gates of the mikvah.

NOTE

1. *Gehinnom* is the valley of eternal hopelessness.

CHAPTER 7

THE SEVENTH SAPLING

One day Rabbi Yohanan was swimming in the Jordan. Resh Lakish saw him and jumped into the Jordan after him. [Rabbi Yohanan] said to him "Your strength is for the Torah." He [Lakish] said to him: "Your beauty is for women." He said to him: "If you repent, I will give you my sister, who is more beautiful than I." He accepted upon himself [to repent]. He wished to return [with a jump] to get his clothing, but he was unable to do so. He taught him Bible and Mishnah and made him a great man.

An interesting meeting between a teacher and potential student, I think. Yohanan, because he is a scholar and a teacher, not unexpectedly tells Lakish who was once a more than competent scholar, that he should devote himself to study and scholarship. In more contemporary terms Yohanan might have informed Lakish that he wasn't working up to his potential and that according to his IQ test scores Lakish should be achieving at much higher levels. As might be expected, Lakish answers with a second century version of "Ahhh, your mother wears army boots." Lakish will have none of Yohanan's flattery and certainly refuses to be seduced by the apparently sterile appeal of scholarship. But Yohanan persists in his wooing, this time by coupling scholarship and sex, and it is this pairing that Lakish apparently cannot resist. This compelling union of the spiritual and the physical is beyond Lakish's capacity to refuse, and though Lakish's instincts urge him to gather his things and run, the allure of scholarship is too great and he "was unable to do so." Thank goodness for the appeal of study, or we might all soon flee the comfort of the academy for the satisfaction of the brothels. Again, I am struck by the Talmudic connection between study and sex; Lakish enrolls under Yohanan's tutelage and acquires both beautiful teacher and even more beautiful wife.

Perhaps this story suggests that the goals of scholarship are so vague or remote, or they appear so immediately purposeless that desire to enter into such activity develops very slowly. Wispy filaments of smoke ascend from a pile of leaves stacked densely in a mound; someone walks over to the smoldering pile, kneels carefully before it, and places both hands palms down just beyond the rim of the leaves. Turning the head parallel to the ground, the person blows gently in the direction of the smoke. Once, twice, and the airy puffs of smoke transform into a steady stream; heat emanates from the origin of the smoke stream. The person takes note of the increasing density of the smoke and then leans in again to puff once, twice, thrice more. Rocking back the body waits tensely; suddenly, where once only curled smoke existed, flames burst forth. The mound of leaves has ignited. Much physical effort has been expended, and there is immediate pleasure in the heat and light achieved.

I am myself indifferent honest and was early seduced to scholarship by Raymond Matienzo, as I have said. And perhaps typically, I remember bursting into flame and not the prior effort of being carefully blown upon. I recall, too, being set aflame in the classroom of Mr. John Bartul, my high school American history teacher, though I cannot recall the longer process of smoldering. Mr. Bartul was tall and thin and deeply in love with America and American history, and perhaps with us. The day following the complex weekend of the assassination and interment of John F. Kennedy, John Bartul stood solemnly in front of the room, and reached upwards tentatively with his index and middle finger and touched the flag hanging from the classroom wall. "I have been trying to teach you all year what this symbol means," he said in a voice breaking with emotion. "The events of the past few days have been far more eloquent than I could ever be." And then he paused, obviously overcome with emotion and not wanting to shed tears. He swallowed hard and said, "I can't go on today." And then Mr. John Bartul sat down quietly at his desk and stared out at us, though I am certain he did not see us. I had never seen a teacher express so personal emotion. I was in awe and somewhat embarrassed. I didn't know that knowledge could feel.

I remember clearly also one of Mr. Bartul's assignments. "Re-read chapter 16 in your textbooks." I don't now recall what the subject matter of that chapter might have been; I know now it doesn't matter. He continued, striding across the room with incredible energy: "Then I would like you to write a paper of five hundred to a thousand words in which you point out some of the text's prejudices." I recall being rather puzzled at his direction: how could the text be anything but sacred and perfect and inviolable? How could I, a mere high school junior, discern signs of . . . well, prejudice in the renowned author of my American History text, the esteemed Thomas Bailey? If *his* was a story tainted with biases, an inclining not to be admired, then where might I ultimately go to discover the politically correct American history? Whose account could possibly be more accurate? Didn't I as a student—my goodness, as an American, have to know accurately and finally what happened and to whom? In the recesses of my reason, I felt an alarming concern, a certain, persistent and aggravating gnawing: what, I feared, if there was no correct story? Knowledge, I had elsewhere learned, was contained in these texts; knowledge, I had elsewhere learned, was objective and true and eternal. Suddenly, my certainties were equivocal. I wanted to jump back and retrieve my clothes, but I was unable to do so. John Bartul had seduced me, and I liked it.

Now mostly it is not requisite that we acquire our students, as perhaps Yohanan acquires Resh Lakish. Usually I enter the classroom on the first day of class and the students are already sitting quietly in the same seats they will occupy for the entire semester. Once, during the early part of a semester the class was re-assigned a room from one with an east-west orientation to one on the opposite side of the building with a north-south bearing. My students calmly picked up their books and pencils and overcoats, walked quietly to the new classroom and sat down in exactly the same positions as they had occupied in the previous room! They had been given the opportunity to change their perspectives and yet they chose the familiar. I often wonder if our students are afraid we would not find them if they moved their seats; or that they would lose some spiritual perspective on education by changing their physical point of view; I wonder whether our orderly and linear seat arrangements predisposes students to sit so regularly.

But I think I work assiduously, like Yohanan, seducing my students to become, as it were, my platonic lovers. I invite them to share my metaphoric bed and stay up all night engaged in pleasurable and passionate, intellectual activity. I would, I suppose, be found asleep with them in the fourth hour at the tavern. I want to awaken their senses and stimulate their imaginations in the exercise of intellectual questings. I desire to give them satisfaction that they would become enraptured by the allure of study and become enthralled with it. I read them wonderful love poetry written to seduce so that they would fall in love with me and with the poetry. I tell them: teachers are intellectuals, and to be so requires dedication and hard work and great effort; I expose my intellect and flutter my eyes. I romanticize the struggle in which they should engage themselves at my urging; I seduce them with the discourse of the academy and lure them into the clean sea of abstract thought. I tell them that teachers ought to be intellectuals and that to be so requires intense training. I tell them their beauty should be for Torah. They tell me my mother wears army boots.

Our teachers are our ancestry; scholars often define themselves by those with whom they have studied. We develop lineages quite unabashedly in academia. It legitimizes us to carry our teachers on our backs, as it were. Indeed, Talmud insists that carrying our teachers takes precedence over carrying our parents! Perhaps our scholarly family trees situate us as individuals in The Field—their heritage roots us as particular trees in a specific forest. The reputation of the teacher ennobles the student. We say, "I studied with . . ." or we boast, "My dissertation advisor is . . ." We announce, "I took a class with . . ." One's reputation, it would seem, often begins with the character of the parent. I think Althusser (1971) referred to this as *interpellation*—a calling. Thus, Althusser notes, before the child is ever born, it is given character by the desire of the parents. As for myself, I have myself always felt orphaned in the academic world. I did my graduate work at a commuter school and in an alternative doctoral program that contained no brilliant stars. And then I moved away from my first field, literature, and became a professor of education probably because I forever intended to teach students and not books. I studied mostly by myself. When genealogies are drawn up (William Schubert once produced a genealogy of the Reconceptualist movement for an article in *JCT*. I have always considered myself part of this movement, as I had perhaps felt immediately at home in the existentialism to which Mr. Matienzo introduced me. I published in their journal! But alas, in Schubert's article I was listed neither as student nor teacher on that family tree!), I am regularly confronted by my lack of lineage. As a result, I am usually recorded as absent. I say the aforementioned in some vague defense of myself, I know. But I do often wonder what it is I might have missed having been an intellectual orphan. For me, perhaps, this condition has inspired a certain insecurity, a certain illegitimacy to my claim to scholarship and intellectual authority. No one ever said, "Your beauty should be for Torah," and invited me to study.

Of course, as an orphan I choose my own parents and choose them freely and often. Unlike Jay Gatsby, I have never thought of myself as the son of God, but as more humbly descended from whomever I am reading that permits me to become a character. Perhaps that is why my reading seems so undisciplined: a novel, a work of philosophy, a book on postmodern ethics followed by one on Islamic art and a Harry Potter adventure. What vague longing do I follow to become who I would become? I wonder, what can be said of the descendants of intellectual orphans?

✡

Now, Resh Lakish is considered one of the greatest of the first generation Palestinian Amoraim. These are the Rabbis of the third to the sixth century whose discussions of the Mishnah became the core of the Talmud. It was their conversations that began the process that will finally be redacted into what we know as the Talmud. Lakish was active during the years 250–90 C.E. It is reported that as a youth Lakish had studied Torah and had even gathered a significant reputation as a scholar, but because there was little financial remuneration to scholarship (a condition that remains, alas, true—at least for me), Lakish had sold himself as a gladiator in the Roman arena and became there a successful virtuoso. Some say he turned to robbery as a highwayman. In this interesting story the Rabbis recount Lakish's return to the academy, as it were. It is a result of this chance meeting with the beautiful Yohanan that inspires Lakish to return to the life of the intellect and Torah study. There is, of course, a value judgment in this story regarding the appeals of scholarship, but then, what else would we expect from Rabbinic scholars but an appeal to the beauty of erudition. I mean, these Rabbis invented textual scholarship. There is, though, that puzzling sentence: "He [Lakish] wished to return [with a jump] to get his clothing, but he was unable to do so." I wonder if Lakish had some suspicion what the commitments to scholarship would entail and he was wary of that responsibility. Perhaps our students know as well the obligations inherent in scholarship and they, too, are reticent to make the jump. The appeal of Torah study puts a return to material gain on the bank and just out of reach; study will render Lakish incapable of returning to his former life. It is not a simple decision to change the flow of one's life. In the classroom we avoid this dilemma by reducing scholarship to test answers and achievable objectives. It is not the process we offer (Your beauty should be for Torah!) but the product (the final exam will count as fifty percent of your grade, or I'll give you my sister!). The end of the semester is near, we exclaim out of breath, and we are not yet even up to World War I! Hurry, it won't be here next year!

I love that every edition of Talmud (indeed, I have read recently it is true of every Jewish text!) begins on page two; the Rabbis say that no matter how much we may study, we have not even gotten to the first page. Is there any wonder Lakish wished to return to get his clothing; this decision to become a student would, indeed, change his life. It surprises me that Lakish's opportunity to engage in scholarship occurred from a chance meeting with a scholar at the swimming hole. So much of life—even in academia—seems based in contingency. Another great scholar, Henry David Thoreau, asks, how many people can date a new era in their life from the reading of a book? This story in Talmud attributes such a momentous change to a chance meeting with a teacher while swimming on a hot day in the cool Jordan River.

✡

One year at the high school at which I taught, I was assigned an eleventh grade honors' English class. This was an upper middle class community with highly motivated and ambitious students. This was a school in which almost everyone would go on to some post-secondary education, and the authority of the teacher derived from the emphasis placed on good grades and high Scholastic Aptitude Test scores. I was never quite at ease: what scores students received on the all-important standardized tests was finally attributable to the quality of my work. Could I ever live with myself for denying anyone the opportunity to attend Harvard or Princeton or Brown? I was often intimidated by what I knew I did not know: once a student begged for an extension on a paper because, he said, he was typing a report for the Westinghouse Scholarship, and he didn't want to pull the paper out of the typewriter to begin the paper for my class. These were the days before the advent of computers, when Jeffrey might have simply opened a second document. I offered him the extension readily and casually asked him what project he was working on for the science competition. He told me the title of the paper (it was as lengthy and ponderous as many a paper I have composed), and I didn't understand a word he spoke. This particular English class was, of course, a delight. No matter what I assigned (Madeleine Grumet reminds me that we always assign the books we want to read with others—and that coincidentally happen to be in the school book room), students would read it. I hoped, of course, to seduce them with my literature selections; they were often effusive in their responses. I was flattered. I wanted them to love literature as I loved literature. I wanted them to love the books I loved. Did I think as a result that I, too, would be loved? Once, a lovely girl, Laura, read Henry James's *Portrait of a Lady* and filled two large notebooks with her considerations and reflections on the novel. She then independently wrote a paper analyzing the use of color in James's novel, a paper she rewrote a novel's worth of times. I think she eventually went on to earn her Ph.D. in English Literature. I think of this class fondly as a bacchanalia of literary proportions.

But it is not of Laura or of the entire class that I wish to speak. It is of Daniel Schwartz. He was a pesky, obsessive learner. He had to always be certain that he had heard the discussion or the assignment correctly—after each and every class Daniel would bound up to me and pursue some issue raised either in our discussion or by my assignment. I remember, I am sorry to say, none of his conversation. Perhaps I imagined he had already been seduced; my attentions were elsewhere required. More probably I was not attracted to him. He seemed to me in a constant state of disarray, papers always leaking out of his loose-leaf binder, his textbooks stacked askew under his left arm, and a ballpoint pen poised always ready in his right hand. His shirttails always hung out of his trousers flouncing carelessly as he surged about. Almost all of Daniel's life was consumed with study—as did so many others there, he wished to attend an Ivy League School; he eventually attended an Ivy League school, as did so many others there. He seemed to have no other interest in life save study: he played on no sport teams that I was aware of, he did not date as far as I could tell and he belonged only to the Science and the Chess Clubs. These are the clichéd extracurricular activities of the Uncool, the geeks, and, in 1984 still, those who yet strapped slide rules to their belts. Perhaps the Rabbis would have found him attractive, but I was at the time bored by my student and not at all attracted to him.

Nine years later I walked into the Bottom Line, a very popular music venue in Greenwich Village, New York. That night *Fast Folk* magazine was holding a concert of some of the musicians who supported and were supported by the publication. As I twisted toward my table in an already crowded room, a voice hailed me, "Mr. Block! Mr. Block, it's me, Daniel Schwartz. Hey, how are you?" Indeed, it was Daniel Schwartz, a bit older and thinner than the student I remembered, and his acne had all cleared, but certainly this was still the smile I knew from the classroom. "Gee, I'm fine, Daniel. And how are you? Its nice to see you here." I made some inane conversation; I don't remember having much to say to Daniel in high school, and I did not yet know that a new era in his life had begun and he was not still what I thought him to be. During the small talk in which I struggled to engage, Daniel explained that his presence at the Bottom Line was professional: "I produced the recordings of some of the people in this show," he informed me. I was taken aback. I don't know anyone remotely attached to even minor stardom; even when I am first on line to purchase tickets, I never sit closer than the rear of the auditorium. I was once the first person to call for tickets for Arlo Guthrie's annual appearance at Carnegie Hall and I still sat in the back row. Now Daniel was telling me that he had produced the albums of the people whose records I had purchased and whose performance I was now attending. "I have a recording studio at my home in Mill Creek. I've been doing that work for a while." He continued. "You know, Mr. Block, I wanted to thank you. If it hadn't have been for you, I might have remained a terribly limited person. You taught me there was more to life than books."

That was a shock. I taught only from books, or so I thought. But I was flattered and not a little proud. Then I asked, "Well, thank you, Daniel," I offered. "Is this what you studied in school? Music production? Are you involved professionally in the music industry?" And he calmly responded, "Oh no, I do this for pleasure and fun." "Well, what did you study in school, Daniel? What do you do?" I asked not a bit condescendingly. "Oh," he said, "last week I graduated from medical school."

I was stunned. Not that Daniel had graduated medical school; a professional education was *de rigueur* for the students in these honor classes in Mill Creek. So many of them are now doctors and lawyers, architects, and engineers and scientists. No, what overwhelmed me was Daniel's offhanded revelation of what remains a remarkable achievement. Indeed, if I hadn't asked Daniel about his activities, he would never have told me of his accomplishment. Rather, it was everything else that he had learned with me that was important to him; everything I was to him but remained unbeknownst to me. I had had no idea. I know I congratulated him; but I must admit to experiencing a bit of confusion. What did I think I had been doing all that time in the classroom? What happened to all that scholarly activity? Well, I thought, so much for my carefully wrought lesson plans!

✡

One day there was a difference of opinion in the study hall. "A sword, and a knife, and a dagger, and a spear, and a handsaw, and a sickle—from when are they susceptible to ritual impurity? From the time that their manufacture is completed? And when is their manufacture completed? Rabbi Yohanan says: From when he tempers them in the furnace. Resh Lakish said: From when he furbished them in water. [Rabbi Yohanan] said to him: "A robber understands about robbery." He said to him" "And what good have you done me? There they called me 'Master,' [and] here they call me 'Master.' He said, "I have done you good by bringing you under the wings of the Divine Presence."

Yohanan was upset that on a certain legal matter his student, Lakish, would not only develop ideas for himself, but publicly offer an opinion differing from that of his teacher. When Lakish disagreed with Yohanan on an issue of law, the teacher countered by reminding Lakish of the student's less-than-glorious past, and of the painstaking and glorious effort undertaken by the teacher to inaugurate a new era in the life of his student. That is, Yohanan informs Lakish that the teacher's opinion always takes precedence over that of the pupil because the teacher remains always the teacher and must always rise above the pupil's past to which the student must be forever bound. Of course, Yohanan errs by thinking that Lakish's past can be of no service to his scholarship in either the present or the future. Yohanan errs by thinking that Lakish's past could not help him know anything his teacher has not explicitly taught him. Yohanan errs by thinking that only his teaching instructs Lakish. Yohanan errs by demanding that the teacher's authority derives from his social position rather than from his wisdom. Teachers say that they love when their students disagree with them, but there may be a certain speciousness to this assertion; perhaps I have limits to the degree of disagreement I will tolerate. Perhaps we want our students to develop their intelligences, but in that flowering not to rival that of their teachers—at least until after they have left our class and as long as they do not return to disagree with us. We may tolerate other blooms so long as we remain the tallest growth in the garden.

Lakish, however, refuses to be humbled by Yohanan's stern rebuke. Indeed, Lakish reminds Yohanan, that in his other life as gladiator and as robber, he had been referred to as "Master;" Yohanan has offered little to Lakish whose learning has earned him (again) the honorific "Master." There has not been much change, Lakish argues, in my status in the world as a result of Yohanan's teaching; rather, I have merely changed fields. Of course, Lakish is perhaps, intentionally, provoking his teacher; but then, his teacher is denying Lakish the very ability he has been studying to achieve—the capacity to think independently. Perhaps, it was not finally the ability to think that Yohanan sought to inspire in his student (and brother-in-law), but the willingness of his disciple to attend and admire. It is not under the wing of the divine presence that Yohanan would have Lakish; rather, it is under his own arm that he has gathered him and Yohanan evinces a formidable and protective jealousy of his control. But if Lakish's presence under the Divine Presence can be attributed to Yohanan, then is Lakish bound to unquestioning fealty and silence as his mentor's payment? If we influence those in our classrooms to be students, then of what does their obligation to us consist? Lakish's success is

Yohanan's problem: having taught his student to question, he cannot control Lakish's questions or answers. It were better, Yohanan perhaps considers, had his student not been saved.

✡

Rabbi Yohanan was deeply offended [and] Resh Lakish became ill. His [Rabbi Yohanan's] sister came [and] wept and said to him: "Act for the sake of my children." He said to her: "Leave your orphans, I will preserve them alive.'" "Act for sake of my widowhood!" He said to her: "And let your widows trust in me.'" Rabbi Shimon ben Lakish died, and Rabbi Yohanan was greatly distressed about him. The Rabbis said, "Who will go [and] relieve his mind? Let Rabbi Elazar ben Pedat go, for his statements are sharp." He went [and] sat before him. [After] everything that Rabbi Yohanan said, he said to him: "There is a Baraita that supports you." He said, "Are you like the son of Lakish? When I would say something, the son of Lakish would raise twenty-four objections against me, and I would give him twenty-four answers, and the statement would thereby be clarified. And you say: 'There is a Baraita that supports you.' Do I not know that what I have said is right?" He went on rending his clothes and weeping, and said, "Where are you, son of Lakish? Where are you, son of Lakish? And he cried out until his mind slipped from him. The Rabbis pleaded for mercy on his behalf and he died.

Indeed, Yohanan's intellectual arrogance and personal forwardness renders him incapable of human compassion. Insulted by Lakish's independence, Yohanan haughtily breaks with his student who then falls seriously ill as a result of this fissure with his teacher. Yohanan's sister, whom Yohanan had married to Lakish as reward for his engagement in the study of Torah, begs her brother to make amends with her husband and Yohanan's brother-in-law. But Yohanan's pride will have none of this harmony. He abjures any responsibility towards Lakish: Yohanan expresses what will come to be known in later centuries as Faustian pride. He will be, he announces to his sister, all that she will need for sustenance. He will replace Lakish as the children's father. His insistence that his opinion is unquestionable and his refusal to brook an alternative to his scholarship destroys the Other. If Lakish will not be Yohanan, then Lakish might disappear. If Yohanan must acknowledge Lakish's independence, then it were just as well Lakish not survive. And indeed, Lakish dies. I think the story suggests that Yohanan is, indeed, responsible for the death of Lakish with whom he will not make amends. The teacher makes the student disappear in his refusal to acknowledge him. Yohanan destroys Lakish by his denial of his ability to have a life—to have thought—independent of that of the teacher.

No sooner does Yohanan lose his student, than he mourns Lakish's loss and falls into what we would today refer to as a deep depression. In the absence of Prozac, the scholarly Rabbis advocate that relief ought to come in the form of intellectual disquisition. How better to revive Yohanan's failed spirits than engagement in scholarly conversation! The Rabbis send their best scholar, Elazar ben Pedat, who listens in awe to Yohanan's brilliant discourses. But Yohanan is not assuaged; the conversation bores him. He accuses Elazar: What good are you? Don't I know I'm right? I need you to tell me I'm wrong. You must argue with me, Yohanan scolds Elazar, you must contradict me that I might continue to learn. How can I be a student, the teacher asks the student, if you, the student, will not be the teacher? Yohanan has learned that the student's value is not in agreeing with the teacher but in disagreement. Lakish's brilliance rests not only in what he learned but in what he could help Yohanan learn. Yohanan has lost his best teacher and his best student when Lakish dies. Yohanan cannot go on.

Thoreau had asked how many men can date a new era in their life from the reading of a book; I have asked how many men can date a new era in their life from the meeting with a teacher. The Talmud seems to ask how many men can date a new era in their life from the meeting with a student.

The teacher who is not challenged cannot learn; the student who will not challenge needs no teacher. All authority is invested in the relationship founded on query and response. The authority of the teacher exists in the authority of the student; the authority of both resides in the text that is opened by query and question.

✡

And even so Rabbi Elazar the son of Rabbi Shimon did not rely on his [own] opinion, [and] he accepted upon himself afflictions. In the evening they would spread for him sixty felt mattresses, [and] in the morning they would draw from under him sixty basins of blood and pus. The next day his wife would prepare for him sixty kinds of food made from figs and he ate them and recovered. But his wife would not allow him to go out to the study hall, so that the Rabbis would not push [and hurt] him. In the evening he would say to them [his afflictions]: "Come, my brethren and my friends." In the morning he would say to them: "Go, because of neglect of Torah."

Poor Rabbi Elazar. He can't seem to get over his own doubts; every evening he makes space for the suffering he must endure because he condemned a man to death. Perhaps Elazar is emblematic of all of us who would be certain in our roles as teacher, parent, or lover but cannot, alas, be so. We are oppressed by our doubts. Despite the assurances of his students that the man whom he condemned to death (who referred to Elazar as "Vinegar, Son of Wine") was, indeed, a wicked man, Elazar cannot put himself at peace with the role he played in the man's execution. And despite the fact that Elazar's fat did not decay in the hot summer's sun despite signs of blood marbling throughout it, Elazar remains haunted by the consequences of his accusation. He willingly suffers over his culpability. I am fascinated by his degree of personal responsibility and sense of culpability.

Once there was a student. A young boy, this student loved to play and hated to work. School was to him work. Though he would sit in class, he was always in motion. So much energy comprised this child that his pen could never stay long on the paper, and though it would start ably enough, by the middle of the line that pen had moved into flight and the sentence never reached its destination. But then, he was young and did not care about the future or getting his thoughts down articulately on paper. And yet, his teacher was obsessed by it. Terrified that without constant vigilance, the teacher would despoil the future by the failure to properly prepare its inhabitants for its momentous arrival, the teacher would conscientiously pull the boy through the curriculum as one might pull a young child into the bathroom to wash hands and face.

But the boy smiled so beautifully that his teacher loved him. In that smile, the teacher saw an innocence that he cherished in the world, even saw the innocence he would seriously taint. When the student would not study for the quizzes and the tests the teacher prepared in order to evaluate the quantity of material the student had retained, the teacher would give this student a failing grade; and he would smile at the student with concerned affection. He would gently admonish the student with words of love employing the vocabulary words the student had still to learn. When the student failed to turn in his writing lessons, the teacher would keep him after school and sit next to him and help push the pen across the paper. The teacher hoped that his concern for this student evidenced in the proffered smiles and failing grades would seduce the student to greater effort, and then the student would succeed. The teacher hoped that his own passion would become the beacon that the student might use to steer safely to shore.

But the student remained, as it were, happily afloat at sea. He sensed no danger. And so the teacher caused winds to blow and waves to pound about the student. He sent home warning notices and called parents in for conferences. After class he would hail the student to the front of the room to urge him to make greater effort. And the student would smile and assure the teacher that yes he would definitely do so.

But the future remained too far away, and the teacher's urgings remained unheeded; the boy continued un-enthralled by the books and words with which the teacher meant to seduce the student. The student did not heed the teacher's beacon. The student did not believe the teacher when he warned of the consequences of his inactions, as it were. Nor did the student fall in love with either the teacher or the subject matter. Or perhaps the teacher could not recognize the love of his student because he could not see the student through the filter of the teacher's own image. And so at the end of the year the teacher failed the student with a grade point average of fifty-seven. Sixty was a passing score.

And the student went to summer school for the three points.

This morality tale has no moral. But its persistence in my memory suggests the doubts I still retain concerning my behavior almost thirty years ago. This remains for me a very painful memory, but I think it serves me well in the present. Academic suicide is today an option in my classes, but not execution. For me now, the issue concerns not the matter of the three points: had I been wiser I would not have failed the student. But the three points seem symptomatic of a larger concern and perhaps illumine a somewhat different issue. If a grade is assigned based on a computed average of a body of quantitatively assessed work, then in the assignment of grades a teacher ought to consider a fifty-seven as inviolable. And whether the computed average is one point or three points or thirty points from passing, it ought to remain a failure. Perhaps few teachers would fail a student with a fifty-seven, except perhaps Mr. Feeny on *Boy Meets World*. Most of us, I believe, would assign a passing grade based in some subjective assessment—effort made, attendance, extra-credit assignments, compassion. But, as soon as I adjust these numerical computations (often via numerical assessments of subjective qualities) then I have de-legitimized any final evaluation based in these numbers. If these computed averages are by nature uncertain and their meanings so ambiguous that they can be adjusted by subjectivity, then how can we ever depend on any computed average for a permanent assignment of final grades? If my subjective assessment can adjust the objective evaluation, then of what value is the objective assessment? If my subjective assessment can alter the outcomes of the objective evaluation, then what role does my subjectivity play in the construction in what we think of as objective assessment? Is a B+ an 87 or an 88? If I cannot tell the difference between a C and a C+, then how can I tell the difference between an A and a C? And if I can't be assured of the difference, then how can I sentence anyone to either grade?

And finally, I know that in a basketball game three points might mean the difference between winning and losing, but in the classroom where the final buzzer will not sound for years if at all, save at death, could three points have any value whatsoever?

✡

I am intrigued as well by the Rabbis' return in their discussion to Elazar's guilt. Elazar, we are told, continues to willingly suffer, and the Rabbis cannot cease considering their colleague, Rabbi Elazar: despite the meanderings of the conversation in Talmud, the Rabbis keep returning to Elazar's story. Clearly, his actions trouble them. Perhaps what is so interesting about him is the finality of his acts: his word condemned another to death. And now he is beset by uncertainty and doubt. The redactors of the Talmud themselves continue to be troubled by their reservations concerning certainty: *it is not in Heaven.* They are justifiably troubled by Elazar. Doesn't their anxiety derive from the same place as mine: the absence of certainty? Don't we all suffer because we cannot be certain of the consequences of our actions? As teachers we would love to achieve certainty in our knowledge and our methods, and so we continue to invent mechanisms to evaluate and mark progress as if learning were merely a journey on a clearly marked and detailed path. I think we sometimes forget that what we teach is what we have ourselves learned over a quantity of years, which often exceeds the age of our students. They haven't yet had the luxury of time we have enjoyed.

Nonetheless, every morning Elazar orders the suffering to be gone and returns to his holy studies. As must we all get on with our duties. And so we all arise and banish our doubts.

✡

One day his wife heard [and said] to him: "You bring them upon yourself; you have squandered the money of my father's house!" She rebelled [and] went to her father's house. Sixty sailors came, [and] brought him sixty slaves bearing sixty moneybags, and they prepared for him sixty kinds of food made from figs, and he ate them. One day she said to her daughter, "Go [and] check on your father [and see] how he is doing now." She came, [and] he said to her: "Go [and] tell your mother: 'Ours is greater than theirs.'" He quoted about himself: "She is like the merchant ships, she brings food from afar." He ate and drank and recovered, [and] went out to the study hall.

I wonder what the meaning the number sixty has to the Rabbis: sixty sailors, sixty slaves, sixty moneybags, sixty kinds of foods made from figs! I am a bit troubled with their repeated use of it: in school, sixty is a passing grade!

Of course, Elazar's suffering is not accomplished in isolation or without social cost. For the relief of his sufferings, Elazar's wife has expended her entire inheritance and the limits of her patience purchasing potions and medicines and doctor's care. She complains bitterly to her husband: "Your sufferings are all self-inflicted. They have ruined our life. It needn't have been this way." Infuriated by his masochistic and, to her mind, self-absorbed behavior, Elazar's wife disavows her commitment to Elazar and his modes of life and, with her daughter in tow, she flees Elazar's home and returns to her father's house. Perhaps now she is free from his obsession and can get on with her life and the raising of her daughter.

Having spent glorious years of my life in often painful therapy, it has become impossible to consider my activity of the world outside of the therapeutic process. Each of my acts is accompanied by an acute self-consciousness that interrogates the multiple motives and myriad effects of every character stance I assume. Now this self-analysis is usually not a burden to me, though I occasionally find it wearing; certainly, it disturbs my sleep and I arise once in a while to empty my blood, as it were, into the basins. I expose my fat to the heat of Av and Tammuz. But, this reflective consciousness has come to be essential to my existence. If there is the self that acts, then there is also the self who contemplates and reflects upon where the acting self has been and what it has done. This reflective self considers what meaning might be constructed from the acting self's actions. This multiplicity of selves—what Christopher Bollas (1992) refers to as a *complex self* (perhaps to offer a naturalized image of a non-schizophrenic multiplicity)—is forever engaged in a world of objects and forever engaged not only *in the use* of those objects, but also in the interrogation *of the use* of those objects. In such activity, the self continues to become character. Isn't that what Whitman might have meant when he said, "I contain multitudes?" We are an infinite number of selves evoked from our relations with the objects in the world. In therapy, I have learned new elaborative means for understanding (a form of use) and using the objects of this world. Other strategies for object relations I have learned in school. I learned, for example, how not to read a textbook. Sometimes it has been a painful learning that I invited to my bed: I was often not as successful as Elazar in banishing them from me for my periods of study.

Others, however, sometimes display very little patience for this self-reflexivity: they say, It is a bourgeois self-indulgence to understand the world from the perspective of the self; it is tiresome, they complain, to evaluate every nuance of behavior; it is tedious to question every action; it is false, they accuse, to assume that behind every action is a story. I do not always know what they would prefer I do, but these acquaintances evince little patience or understanding for my speculative turn. Oh, they are never rude, though sometimes they behave a bit abrupt. They weary of my self-consciousness. Sometimes they roll their eyes; sometimes they dismiss my interpretation because they prefer a good cigar. Nonetheless, as a part of every thought and act, I require that its context must be painstakingly elaborated and its cathexes defined. It is how I live. A large part of the burden of this self-reflexivity falls upon my children, as you might expect. Every act in which we engage together walks with them, as it were, towards the analyst's chair. When we are ebullient, I hear my children say to their therapists, "My father was always so cheerful. He gave us our *joie de vivre*." When I am snappish I hear my children likewise say, "My father was always so short-tempered. We were always uncertain how any of our behavior would be received." When I am angry I hear my children tell their therapist, "My father was always angry and I think that results in our own fearfulness." When they sit with their books I hear them say, "My father loved books. I think we get our love of learning from him." With one eye I watch my activity and with the other watch my children write the check to the therapist. I myself write the check to the therapist. I do not know how there is any relief from this. Certainly the story of Adam and Eve suggests there is no return to the Garden and no forgetting our nakedness. Once they saw evil, they could never again see the good without its admixture of evil.

As this is how I live, so is it how I teach. There is no forgetting that I am a teacher. As there is no forgetting the enterprises of the therapist's chair, neither is there a forgetting of the concerns of the lecturer's podium. It is just not possible to leave that character at the office, as it were. I mean, there are children living here! I am, it would seem, forever engaged in offering to them narratives of the self and the world. I am, it would seem, forever engaged in disrupting the comfort of the narratives they insist are true. I have expended thousands of dollars in the purchase of music, books, and toys to inspire the education of the children by immersing them in the objects that comprise the world I want to introduce to them. The more varied the world to which I aspire for them, the more varied the objects I must bring to them. I have squandered my inheritance and that of my wife.

There is a connection between the classroom and the home. Bill Pinar (1994:257) reminds me of it. He says in "The Lost Language of Cranes," one of my favorite of his essays, that "[P]arenting, like teaching, is an opportunity to recover lost languages, to speak again, even if silently, from one's childhood." I relive my childhood in my children: I think it is my childhood that will lead them ultimately to the therapist's chair. And I also give them something beyond my childhood but about which I cannot speak, and that unarticulated object that I bequeath them derives from the silences that comprised much of my childhood. It is through the discovery of these silences in therapy and therapy-like conversation that I learn to speak a differently informed discourse.

I think that it is these ghostly voicings that speak to the children in the language of the therapist's office. Similarly, too, our children learn from my own engagement with the objects I have learned to study and with which I study. It is with the alternative story, the other discourse, the novel narrative with which I mean to allure them into the world I know. Bill says that it is this seduction that lies at the core of parenting. Don't we all mean to attract our children to our lives—to our selves? "[P]arental love lives on the sublimated side of incest. Sublimated, that's a relief, I agree. But I do think that stripped of the erotic altogether, by that I mean that without a certain teasing, maybe we can even say a certain seductive element, the parental posture risks becoming merely power-seeking and power-wielding, at times, good old-fashioned authoritarianism" (Pinar, 1994, 258). If the world I present is not attractive, if the children do not feel that the world(s) that they grasp are those of their own choosing, if the children do not see opportunity in the worlds I offer, then I cannot expect my children to desire to enter into them with me. I could, however, force them to do so—but at very great cost. I could squander my entire inheritance.

So too, in the classroom. I think it is my own obsession with education that is evoked in there. It is with my own passionate faith in learning and study with which I mean to seduce my students; it is the sensual delights of inquiry and discourse and the promise of continuous satisfaction in this engagement with which I tease my students. There is a Hasidic tale:

Once a new Torah Scroll was being dedicated in the House of Prayer. Rabbi David Moshe held it in his hand and rejoiced in it. But since it was large and obviously very heavy, one of his Hasidim went up to him and wanted to relieve him of it. "Once you hold it," said the Rabbi, "it isn't heavy any more" (Buber 1947).

By my joyful embrace of arduous study, I hope to seduce others to carry the weight. I would have my students appreciate, even love, the stories I tell and the lives I present. I would have my students long to tell their stories and the lives they would present. I would inspire my student's passion for study from my own intense desire for it so that the stories of their lives would be willingly told and worth the telling. Perhaps the good teacher, like the good parent, seduces with love and, it is hoped, with sexuality sublimated.

I think that our lives are a story; I have said repeatedly that curriculum is the story we tell our children. The idea of the grand narrative has been banished in our time; I am comfortable with that act of social refusal. I have myself in various ways been rendered absent from that master narrative anyways. But I think the Rabbis long ago made legitimate alternative stories. Talmud teaches that interpretation is not in Heaven and that all may be interpreted according to local custom. Storytelling is a human prerogative. The richness of our world derives from the complexities of the stories we can tell about it: I want my children to have the resources to purchase that richness. I want my children to understand the narrative reality of their lives. I burden them with questions for which the story is an answer. But in the schools today it is not the questions that have become standard but the answers. Students demand not questions but answers. It is the former from which the story arises, however. The Haggadah teaches that whoever *enlarges* upon the story of the Exodus deserves praise. But when I offer students the plenitude of stories, they bemoan the absence of the answer. They do not wish to be confused by alternatives. We willingly complicate our lives by what we come to know and the connections we are taught to make. And we willingly restrict our lives by denying the other story. To be a teacher is to know that there is never an end to learning, that there is always another story that could be told and that if I look with care I might discover the existence of that other story to tell. Learning teaches how the stories develop and what they might mean. And for that maturity we require experience: facility with communicative systems and communicants.

Now I use the word *maturity* with caution: Since I have been an adult I have never permitted anyone to refer to me as mature. Kurt Vonnegut (1988) said that maturity was a disease for which laughter is the only cure, if laughter could be said to cure anything. I think he meant that maturity is usually understood as a movement into seriousness and responsibility associated with adulthood and that results in a severe loss of joy. Rather than mature, I would be ironic: an awareness that there is more to every story than is told and that there is another story than the one presently being narrated. Laughter and tolerance should result from such a stance. I once told Emma a joke and she laughed joyously. Several nights later we were sitting at the dinner table and she said, "Dad, I told that memory joke at the sleepover and no one laughed." "I know," I answered, "I know. There will always be good jokes you tell for which you will get no response."

In the story above concerning Elazar, it is said that the sailors had been caught in the storm and had been saved by the invocation of Elazar's name in their prayers. They brought him for his ease sixty slaves bearing sixty moneybags; they prepared for him sixty kinds of food made from figs. He survived despite his wife's desertion. (I am of course reminded of the sailors who tried unsuccessfully to protect Jonah. Unlike Elazar, he had to be thrown overboard). In this Talmudic story, where Elazar's name alone is sufficient to save these endangered sailors, the Rabbis say that the sailors delivered these goods in gratitude to him. I do not like that interpretation at all. For me, there is nothing in the text to support such a reading. Of course, as I am learning, there need not be anything in the text to support a specific reading, and so I prefer to say that since I don't like the interpretation of the Rabbis in question, I will offer my own. I prefer to understand Elazar's good fortune to have derived from his continuous study, which brings the world to him and by which he is then sustained. Slaves, money, and food are delivered to his doorstep, and he recovers from his illness. I am suddenly suspicious that perhaps it might have been Elazar's idea to drive his family away with his suffering so that he could continue to study! And perhaps I have returned to the contradictions that arise in the demands of study and living: has Elazar forced daily life out from his home that he might continue his study? The Rabbi of Kobryn said, "If it were within my power, I should hide everything written by the tzaddikim. For when a man has too much knowledge, his wisdom is apt to be greater than his deeds."

✡

They brought before him [Elazar] sixty kinds of blood, [and] he declared them ritually pure.

The Rabbis murmured [against him] and said: "Can it enter your mind [that] there is not one doubtful [case] among them?"

He said to them: If it is as I [say]. They will all be males; and if not, there will be one female among them." They were all males, and they named them Elazar after his name."

It was taught: Rabbi said: "How much procreation did this wicked [state] prevent in Israel?"

Purity was of immense importance to Biblical Judaism and all bodily discharges were all suspect. Because the command to the people was to be holy (Leviticus 11:44–46), then the maintenance of ritual purity became part of the practice of holiness. In the Bible, this responsibility falls equally on men and women. Judith Hauptman (1997, 148–49) writes that biblical religion regarded genital discharges from men and women alike with revulsion and that the cult of purity transcended gender. Leviticus 15 deals with the subject of the kinds of discharges from the human body that may require offerings as part of the purification process, and the entire first half of the chapter deals with the subject of discharges from the male body that require not only sacrificial offerings, but ritual immersion as well. It is the second half of Leviticus 15 that states the laws regarding the menstrual period. There are, interestingly enough, no requirements in Leviticus requiring immersion following menstrual bleeding or other irregular discharges from the female body; rather, these practices derive from Rabbinic decrees that instituted patterns of control over the female body.

The Rabbis of the Talmudic period altered the parity between men and women in matters of bodily discharges; the control of women's bodies was more assiduously organized by men's rules. Laws controlling menstrual and other uterine bleeding continued to be enforced, while laws dealing with seminal and other male emissions were ignored. Leviticus had required immersion by men to restore purity; the Rabbis had instituted immersion by women. I do not mean to excuse or explain away the obvious sexism implicit in this story; it is, of course, indefensible.

Judaism subsequently developed an elaborate series of practices based in a woman's experience of the flow of blood. The story above concerns the case of women whose bleeding occurred at times other than their regular menstrual cycles. Questions concerning engagement in sexual intercourse as well as other issues of food preparation and household maintenance usually ascribed to women depended on the ritual purity of the discharged blood. Leviticus 15:25 commands that a woman who bleeds from her uterus is ritually impure, and intercourse is forbidden with her husband during these times. However, there might be a type of bleeding that results directly from the reproductive tract and thus, causes no impurity. There appears to be a group of Rabbis whose own experiences may have led them to a certain liberalism in their beliefs and allowed certain leniencies in the categorization of ritual impurity. Since there could be doubt, the Rabbis in the story above sought the opinion of one who could make this difficult distinction. Rabbi Elazar is asked to render judgment on the ritual purity of sixty kinds of bloodstained garments from sixty different women. (Ah, how that number sixty keeps reappearing!) Elazar examined each and declared every single case ritually pure. That is, Elazar passed everyone! Perhaps Elazar's experience as an officer of the King had provided him an understanding of human uncertainty and frailty. Plagued by his own doubts and aware of, finally, how little certain he was of his own knowledge, Elazar refrained from any judgment that would prevent further conjugal relations. I appreciate Elazar's doubts. I myself would rather now pass a doubtful case than condemn a student to failure.

But, as I have come to expect, there is a dispute amongst the Rabbis regarding the reference in the latter part of this story: what is meant by the phrase, "the wicked state"? Some say that it refers to Elazar's sufferings, which were such that he could not attend the Study Hall more frequently and thus could not rule on the purity of more women and thereby enable further procreation. Since it is from the earlier story that we learn that Elazar seems able to banish his sufferings in the morning that he might study Torah, this explanation lacks some plausibility. Some say that "the wicked state" refers to Elazar's wife who would keep him from the Study Hall and therefore prevent him from declaring more women ritually pure and available for intercourse and procreation. Rashi argues—an explanation I prefer—that the phrase refers to Elazar's employment by the Roman authorities, which also kept him from the study hall; his preoccupation with government work would not allow him to rule as often on ritual purity. Thus, argues Rashi, Elazar's rulings would not enable more intercourse in the land, an activity that would fulfill the positive command "To be fruitful and multiply."

And so I think that one moral of this story is that too much attention to national and standardized curricula is a wicked state and inhibits learning.

I think that there may be another plausible explanation for the positive judgment rendered by Elazar concerning the remarkable purity of the women: perhaps Elazar had himself emitted a few bodily fluids in less than pure circumstances. As a result, he gave everyone an A.

✡

> When he [Elazar] was dying, he said to his wife: "I know that the Rabbis are angry with me and [that] they will not look after me properly. Lay me down in the loft and do not be afraid of me."
>
> Rabbi Shmuel bar Nahmani said: "The mother of Rabbi Yonatan told me that the wife of Rabbi Elazar the son of Rabbi Shimon told her: 'Not less than eighteen and not more than twenty-two years I left him lying the loft. Whenever I went up [there], I examined his hair. Whenever a hair fell out, blood would come [out]. One day I saw a worm come out of his ear, [and] I was distressed. He appeared to me in a dream, [and] he said to me: "It is nothing. One day I heard a slight against a Rabbinic scholar and I did not protest as I should have.""'"

I do not think that one would have to have great knowledge of psychological matters in order to recognize the obsession the Rabbis exhibit about their colleague, Rabbi Elazar. I mean, his complicated status in their society is clearly expressed in their compulsion to narrate these bizarre and supernatural stories about him. On a personal level, they don't seem especially fond of Elazar, but they just can't seem to let him rest, as it were. For example, his employment as officer of the king engaged to arrest thieves and malcontents did not ingratiate him to his social or political community. Nonetheless, as we have seen and shall continue to see, there are stories recounting how the community continued to require from Elazar his judgment concerning matters of ritual purity and justice. At some point, the Rabbis excommunicated Elazar for his refusal to accept the opinion of the majority with regard to Torah law, and yet there are stories that sanctify him for his piety and spiritual acumen. His wife, weary of her husband's self-imposed suffering, abandons Elazar and escapes with her daughter to her father's house. Yet this present story recounts how she then devoted between eighteen and twenty-two years (!) keeping watch over and caring for Elazar's body following his death. I'd say these disparate portraits of Elazar in these stories express unresolved conflict in the Rabbis' feelings toward their colleague. In these stories, Elazar is portrayed as a stranger to every aspect of society—social, spiritual, and personal—and yet integrally involved in the operations of each. Elazar operates in each realm bounded by animosity, betrayal, and suspicion, yet he is portrayed in each at its center and serving as a figure of authority. In these stories, despite great pressure to conform, Elazar remains unremitting not only in his behavior, but in his refusal, even in death, to bodily decay. Lying unburied "not less than eighteen and not more than twenty-two years," Elazar's body, the story goes, resists decomposition. Even though he has long been deceased, Elazar's blood continues to flow through his body. He remains, as we shall see, quite vital. These stories represent Elazar as an unswerving standard of intellectual and ethical power who commands an extraordinary degree of interest and even respect. I would like to understand this fascination the Rabbis express toward Elazar.

I would like to account for the apparent contradiction with which the Rabbis regard their colleague, Elazar, son of Shimon. Now, contradictions are the very stuff in which the Talmudic doctors revel. They do not seek so much to resolve them as to show why what appears to be contradictory is not so at all. They maintain the validity of separate opinions by showing not how one is false, but how several may be true. How can the Rabbis testify to their acrimonious dealings with Elazar when in their testimony they portray his sanctity and scholarship?

I look toward the end of the story to justify its beginning. At the end of this episode, Elazar's wife sees a worm coming out of her dead husband's ear. After a significant number of years of her husband lying unburied in the loft, his wife rightly fears the imminent decay of his body. Her husband appears to her in a dream: Elazar tells her that the worm represents another form of suffering that he must endure for having once heard a slight spoken against a scholar and not having protested with the sufficient speed and energy that such calumny should have compelled. I wonder: what a mighty punishment for such a small act! I mean, there is no indication that Elazar was even a principal in the suspect conversation—he might only have overheard the slight. Nor is there an indication that Elazar said *nothing* in the scholar's defense— only that he did not "protest as I should have." I wonder at the severity of his punishment. Perhaps we might hear in Elazar's statement a reflection of the Rabbis' own guilt concerning their conduct towards their difficult colleague, Elazar. In this Talmudic account of Elazar's final suffering, perhaps the Rabbis give evidence of their own contrition for their calumnious treatment of the scholar so unabashedly recounted in these stories. This present story that begins by voicing Elazar's own fears concerning his expectations of the Rabbis' treatment of his body, bespeaks the Rabbis guilt regarding their ill-spoken opinions of Elazar. In giving voice to Elazar's fear, they acknowledge their complicity. They are, as it were, hiding in plain sight. After all, they continued to tell abusive stories about him and even canonized in the Talmud their excommunication of Elazar. But then they proceed to tell tales that sanctify their colleague, Rabbi Elazar.

I think these contradictory narratives give expression to the Rabbis' guilt. Not only did they leave slander unrequited, but they spoke slander as well. The tales told about the ex-communicant, Elazar, paradoxically characterizes the impiety of the Rabbis even as they portray the saintliness of Elazar. It was his fear of their anger that led Elazar to make arrangements for the care of his body after his death; what greater slight to a scholar than the denial of proper burial! In this story, Rabbi Elazar suffers a particular punishment because he left unrequited "a slight against a Rabbinic scholar." What might the Rabbis understand about their own behaviors to tell such a story about Elazar? His fears must have been substantial and real. Elazar was quite fortunate to have a wife of such loyalty.

✡

The gravity, even finality, of the particular act for which Elazar must suffer this final ordeal also intrigues me. Elazar's earlier accusation led to the execution of a laundry man who called him "Vinegar, Son of Wine." The result of this slight was a man's death, a sentence from which there is no reprieve. Subsequent to the execution, Elazar came to doubt his judgment and experienced regret and remorse, though his students tried to reassure their teacher of the legitimacy of his accusation. Indeed, they reassured Elazar that the executed man was a very bad man and deserved his death; he had, they told their teacher, committed unforgivable sins! But Elazar's conscience could not be assuaged. As punishment for his role in the man's death, Elazar invited upon himself sufferings that left him and his family financially and emotionally bankrupt. This was, indeed, a calamitous series of events for all concerned—obviously, we must admit, for the laundryman, but also for Elazar and certainly for his wife and daughter. It interests me, therefore, that it is not for *this* act that the worm eats its way out of Elazar's ear, but for an unrequited slight against a scholar. Despite the fact that in this latter account there are no corroborating students who can justify the slanderous talk directed against the scholar, nor students who will try to rationalize for him Elazar's inaction, the unrequited slight is harshly requited in death. In this tale, there is no relief from this unrequited slander. Here, it is enough that slanderous words were *tolerated* against a scholar to warrant punishment. I think the point is obvious: one must never listen to slander against scholars.

Now, Elazar has been apprehending criminals for the king for years and even seeing to their executions. It is neither a faultless enterprise in which he has been engaged, nor one free from accusations of complicity.

Elazar has himself aimed some condemnatory phrases at a variety of people. Perhaps, he, like Rabbi Yose, might have fled to Laodicea to avoid complicity in the government actions, but that is another story, perhaps, and for another time. Here, however, the Rabbis' obsession with stories that speak of his saintliness belies those Rabbis' belief that this escape should have been taken by Elazar. Apparently, he did well not to remove himself. Intriguing as well is the idea that whereas his earlier suffering derived from his activity, here his suffering derives from his inactivity, from what Elazar *failed* to do. This story responds to his passive stance with an active response: to hear a slight spoken against a scholar and to fail to respond to the slight is a grievous act that demands harsh penance. This category of wrong reaches even beyond death. We know that the Rabbis excommunicated Elazar for not adhering to majority opinion; harsh words must have been spoken in that act. I think that in this present tale, I hear the Rabbis' own contrition.

The scholar, as we might expect from a tale told by a scholar, is afforded special treatment. Even to hear slander of a scholar is a grievous sin. We are, in schools, a community of scholars. Scholarship requires a certain liberality; any hindrance to that openness compromises scholarship. But sometimes it is difficult to hear opinions so widely divergent from my own that their very existence endangers me. In this story I understand that a scholar must not be spoken against but rather, must be spoken to. That is, we must learn not to dismiss but to quest. I think that I would learn to speak as do the Rabbis in Talmud: When an opinion is offered they often begin the conversation with the following question: "From where do we derive this opinion?" Their first response to a difficulty is to query and not argue. I think that the Talmud speaks against academic quietism in this story of Elazar's final suffering. A scholar must not sit passively when another scholar is slighted; finally, it is my own existence I defend in my response. I think the Talmud speaks here of the active value of the engaged scholar. It denies retreat into the academy as insulation from involvement even with those scholars with whom we violently disagree.

Of course, perhaps Rabbi Elazar would be less harsh on himself were he to read the comments which today's scholars direct at their colleagues.

✡

Whenever two people came for a lawsuit they would stand at the door. One would state his case and the other his case. A voice would issue from his loft and would say: "So-and-so, you are liable." "So-and-so, you are exempt."

It is a wonderful image, is it not: two litigants standing at the door to the loft—standing even within the barn door—calling into the darkened interior the details of their dispute to a body they know to be dead yet still lying atop the loft at rest, if not yet at peace. The litigants make their presentations, then stand back, perhaps cross their arms, and patiently await judgment from the deceased. I picture them leaning into the door way and calling tentatively into the darkness. When the first is done presenting his case, the other begins, and the first watches and listens attentively to the presentation of the other. Of course, what else in this situation would there be to look at but the face of the Other? How hard it often is to look into the eyes of those with whom we argue. Both complainants stand quietly and wait for the voice to issue from out of the dark. At the issuance of the disembodied decision, both litigants turn and, perhaps, look at each other, shrug their shoulders, and head for home. Even in death, Elazar's opinions are sought to resolve a dispute. In the previous story, Elazar's voice appears to his wife in a dream; here, his voice emanates from the loft on which his body has been laid to relative rest, and his words carry the power of judgment on certain contentious issues brought specifically to him. This is, indeed, a powerful voice that reaches out into the world of dreams and from the realm of death! The Rabbis wonder, however, how it is that a deceased Rabbi has the authority to decide a local case. The Rabbis have argued that the Torah is not in Heaven, and that law must be explicated here in the local; the Mishnah I have presented began with the decree that all depends on local custom. How, then, could Elazar, the dead Rabbi, decide a local case? An interesting response to this dilemma is that of Zvi Hayyot who argues that as long as Elazar was presented cases dealing only with financial matters his judgments could be valid. His judgments resolve who is liable for payment. Since it is a general principle that a party who is owed money may always waive the rights to that money to which he is entitled, then the one who was declared liable waived the rights to the money s/he had claimed to be originally owed. Thus, in cases brought before him, the deceased Elazar would be making judgments that do not contradict local custom because the waiving of a right is a general principle. Hayyot acknowledges that Elazar would be incapable of ruling on ritual issues, and that is why the Talmud refers to liability—a term that applies solely to financial matters. Hayyot is looking to resolve a real dilemma, and he offers an interesting explanation. Indeed, it is a rather intricate, circuitous one. I am intrigued by his reasoning. It is, I must acknowledge, a remarkable explanation with little justification other than the authority that derives from Hayyot's own wisdom. I think today we still look to the dead though we hold, perhaps, to a different standard of authority.

The dead often maintain power and issue their decrees out of the dark. Elazar's power derives from his wisdom, which extends even into his death. For today's scholar, it is the cry of the footnote and bibliography— voices of the deceased—that authorize the legitimacy of an argument. These departed voices emanate bodiless out of our texts, speaking and authorizing wisdom. In present day academia, we turn regularly to this past to legitimate the present. We use the voices of others to speak our- selves; we define ourselves by the voices of others; we establish our own authority through the authority of others. Footnotes confer authority on the writer! They grow apace on the pages of our texts. Footnotes, Anthony Grafton (1997, 234) writes, "enable historians to make their texts not monologues but conversations, in which modern scholars, their predecessors, and their subjects all take part." It is in the footnote that an author identifies sources of her own material, sources that contradict her own material, and supplementary information that elaborates and clari- fies the sense of the material. Footnotes are an archeological structure atop of which rests precariously the present.

Ironically, the footnote, Grafton continues, guarantees nothing. Rather, the footnote is meant to buttress the argument that the narrator constructs by offering either fervid substantiation or ironic dismissal of material presented in the main text. The footnote locates the work's argu- ment by defining its development and treatment over historical time. The footnotes locate "the production of the work in question in time and space, emphasizing the necessarily limited horizons and opportunities of its author, rather than those of its readers" (Grafton, 1997, 32). Footnotes define how deeply the author has read. There are never enough footnotes. One footnote is too much.

Frank Smith (1990, 135), certainly my candidate for Talmudic scholar par excellence, observes, "[t]he notion [that] 'scholarly' writing can always be tied neatly into a network of other people's publications is academic fantasy. Real life is more complex . . . It is not difficult to find quotations to support any point of view, but nothing is proved by such selections." I think this is more than simply a reworking of the cliché that the Devil can quote scriptures to suit his purposes; what Smith acknowledges, and so much contemporary philosophy and liter- ary theory suggests, is that our lives and our language are part of a web of relations and knowledges so rich and complex that it is fruitless to ascribe any one idea to any one of our sources too exactly. What is pre- ferred is to acknowledge and to hear the myriad of voices in our utter- ances. What we might offer is not citation but autobiography to ground our wisdom. It is the substance of our lives we present as argument albeit, under the guise of another's voice. I ought to wonder not what value there is in my citation of John Dewey, when equally important is the route by which *I* got to John Dewey and what it is that John Dewey means to me. I think our opinions derive not merely from the texts we read, but from the lives we have led. Those lives are constructed by many things, including texts; footnotes are, perhaps, an inadequate structure for elaborating lives. Their crushing weight often collapses the structure they were meant to support.

Those obese Rabbis above made mockery of size: "Theirs is bigger than ours," they cried. They could have been referring to the relative magnitude of the bibliographies of their opponents. We depend on such enormities for our reputations. Elazar, however, doesn't need to provide footnotes to justify his decision nor need he compose an extensive bibliography. Elazar is himself authority.

Elazar requires no footnotes. His authority requires no justification. Like Elazar, footnotes are voices out of the dark. But those voices do not speak for themselves with authority, as does Elazar. Those voices, simply because they are dead, lack power and are manipulable to the purposes of the living. Authority is conferred upon them. There is a difference, I think, between authority and power. Footnotes, perhaps, exercise power. In them reference may be made to an entire text by the mention of a few sentences or disparate words; whole texts are thus misconstrued for lack of a better word. Words of a text are re-contextualized and made to refer to whatever the present writer desires. Sometimes the footnotes are placed at the bottom of the page in a smaller font size; they are reduced to a whisper. Sometimes the footnotes are placed in a special section at the rear of the text itself: they are relegated to another part of the house to which access is difficult. Perhaps placement of these academic structures says a great deal about contemporary scholarship: the citations are intended to justify the conversation but not distract it. In the bibliography, a scholar recounts the readings in which he has engaged in the preparation of this particular volume. They are her autobiography.

It is interesting to speculate whether footnotes evolved because the Talmud was one of the most hated documents of scholarship in recorded history. Indeed, this central production of Rabbinic Judaism, the Talmud, a work of more than a million words, has experienced a terrifying history of castigation, banishment and destruction. For example, in 1240 Pope Gregory IX ordered the burning of copies of the Talmud in Paris; in 1264 Pope Clement ordered its burning; in 1431, a church synod in Basel reaffirmed the stringent ban on the printing and study of the Talmud. In 1553 Pope Julius III ordered the work burned, leading to the destruction of tens of thousands of editions of the Talmud; in 1564 Pope Pius IV announced at the church synod at Trent that the Talmud could be studied but all references which offended Christianity had to be removed. Great sections were excised and some permanently lost. In 1592 a degree of Clement II again prohibited study of the Talmud in any version or edition. Of course, there were other persecutions. It is a long history of oppression and fear that is told. Footnotes are the alternative to the complexity of Talmudic discourse. Perhaps one source of footnotes can be traced to the intense hatred of the Talmudic text. Adin Steinsaltz says that the Sages believed that the Oral tradition—that which was not written—was what makes Jews unique: Talmud rests at the center of this oral tradition. The Talmud is a quintessential model of the importance and centrality of the oral tradition in the practice and development of scholarship. In his historical accounting of the antecedents of the modern footnotes, Grafton (1997, 27) notes that scribes and authors inscribed commentary and annotation directly into the urtext, but he curiously (and characteristically) overlooks the unique scholarship represented in the unique form of Talmud. The Talmud's typesetting looks like what it is: a conversation. It is a form of scholarship. In Talmud, all comment is situated in the body and not at the feet.

A page of Talmud is a unique work of scholarship almost completely unacknowledged in the modern academy. The *sugya* I am studying looks thus: In the middle of the page, set in larger type, are the oldest stages of the conversation: the Mishnah and the Gemara themselves. The first word is enclosed in an ornate frame . . . and then the first Mishnah (the law itself, wholly unfootnoted) is printed out in full. On the page, and set off by the enlarged Hebrew letters gimmel-mem (for Gemara), the Mishnah ends and the Talmudic discussion begins. This is the edited conversation of various Rabbis who lived over the first six centuries of the Common Era. The particular form permits the conversation to occur over hundreds of years. "Rabbi Zera expounded and some say Rav Yosef taught . . ." Who are these people? When did they live and expound and teach? Did they even know each other personally? Talmud does not indicate more than their names. Whatever credibility they maintain derives from their inclusion in the conversation concerning the issues at hand. The issues at hand range far and wide, as is evidenced from our present sugya. You will recall that we began with a discussion of culinary rights of the worker, and we are now engaged in listening to the voice of a deceased Rabbi issuing decisions about (we think) financial matters after he has passed judgment on the ritual purity of the discharge of blood visible on sixty pairs of women's undergarments.

One difficulty studying Talmud arises because often it is not clear which Rabbi is speaking and from which century; another difficulty concerns the anonymous editor. The multi-voiced conversation of this sugya continues in the Soncino edition for fifteen tightly-printed pages and includes numerous and wide-ranging digressions. Steinsaltz (1976, 4) says that

> The Talmud is the repository of thousands of years of Jewish wisdom, and the oral law, which is as ancient and significant as the written law (the Torah) . . . It is a conglomerate of law, legend, and philosophy, a blend of unique logic and shrewd pragmatism, of history and science, anecdotes and humor. It is a collection of paradoxes: its framework is orderly and logical, every word and term subjected to meticulous editing, completed centuries after the actual work of composition came to an end; yet it is still based on free association, on a harnessing together of diverse ideas reminiscent of the modern stream-of-consciousness novel. The Talmud is the edited thoughts and sayings of many scholars over a very long period of time over an enormous range of topics submitted to them in the experience of daily life. And because the Talmud is concerned with the exigencies of daily life the sages are [often] quoted in the present. [T]he work is not merely a record of the opinions of the scholars of past ages, and it should not be judged by historical criteria.

That is, the idea of the footnote is antithetical to the structure and content of Talmud. Rather, Talmud attempts to not distinguish past from present. Rather, Talmud desires to understand both in terms of the other. The past informs the present, but the present also informs the past.

Typeset around the basic Talmudic text, itself already a conversation, is the commentary of the great scholar, Rashi (1040–1105 C.E.). Set on the page outside of Rashi's commentary is that of the Tosafot, which originally were commentary by Rashi's own disciples and students. These are a remarkable set of questions and discussions attempting to connect the text itself with discussions elsewhere in Talmud. Also along the side of the page, numerous other voices from across centuries enter the discussion, cross references to other Talmudic sources are placed, as is a key to quotations from the Bible and another to the great codes of Jewish Law. Additional commentary from scholars in the medieval and more recent centuries is also placed in the margins surrounding the central Mishnah and Gemara.

There has been no book other than Talmud that looks like Talmud.

I wonder if I might model a class after a page of Talmud. What would that class look like?

I would foster conversation and not footnotes. Footnotes lead me away from the present text; Talmudic conversation draws me into it.

Many of my academic conversations are punctuated with the query, "Have you read so and so?" We all have our favorite books. We read the favorites of our favorites. We read the favorites of our favorite colleagues. But it is, I think, an off-putting question. It assumes there is something crucial I ought to, but as yet do not, know. In the question I can hear the retort, "Vinegar, son of Wine!" There is always a book I haven't read; there is always the book out there that I must read; there is always the book out there that will answer all of my questions and solve all of my problems. I keep reading to discover it. Of course, it is a relief not to find it: what would I do tomorrow if I discover paradise today?

Why do we read? There is a glib answer. I will give it to you now: We read for a variety of reasons too innumerable to list. Teachers tell us that we read to learn; we read to gather information. We read for enjoyment. Oh, there is no end to the staid responses to the interminable query. And so I will ask a different question. Here it is: who am I when I have finished *this* book? And let me offer a different and more complex answer: when I read, I gather everything I know from the past and employ it to engage in an activity in the present in order to move toward the future. When I read, I embark on the creation of something unique; no one else could read like me; no one else could read this exact book, no one else could read this book exactly like me. When I am finished with this reading, I should be changed, or how could I account for the time spent reading? It is finally not the answer to the teacher's questions that I seek in reading, nor even the answers to any of the world's innumerable problems. I seek myself in the reading.

In this story above, the litigants must, of necessity, address a dead man: they themselves don't read and do not have the resources for settling the issue. They are stuck in their past knowledges and their past selves. But when I read I construct the person reading and in that activity I must expect change to occur. I do not read to escape the world; I read to construct it. Every book that I read (this only is reading, says Thoreau, that which makes you stand on tip-toe) changes me, and it is requisite that I make some attempt to understand that change. If I must rely on the kindnesses of strangers to assess my reading (and the reading self), then school is a dangerous arena for risk-taking. There are only certain selves permissible within the confines of the classroom, and it is the teacher's questions that define those selves.

Talmud, on the other hand remains an unfinished work; though it can never be completed, it is yet necessary for each student of it to add her voice to it.

CHAPTER 8

THE EIGHTH SAPLING

One day his [Elazar's] wife was quarreling with a neighbor, [who] said to her: "May she be like her husband who was not brought to burial!" The Rabbis said: "[If it has gone] this far, it is certainly not fitting behavior!" There are [some] who say: Rabbi Shimon ben Yohai appeared to them [the Rabbis] in a dream, [and] he said to them: "I have one dove among you, and you do not wish to bring it to me." The Rabbis went to attend to him, [but] the people of Akhbariya did not allow [them], for all the years that Rabbi Elazar the son of Rabbi Shimon was sleeping in his loft, no wild beast came to their town. One day—it was the Day of Atonement—they were occupied, [so] the Rabbis sent to the people of Birei and they took out his bier, and they brought it to his father's cave. They found a snake encircling the cave. They said to it: "Snake, snake, open your mouth and let the son enter to [join] his father!" It opened his mouth for them.

I think I really like Elazar's wife. To my mind these stories and subsequent ones, though ostensibly about Elazar are, in fact, oblique tales told in homage to her. The Rabbis seem to want to portray her as contrary and ill tempered, always fighting with her husband or with others, but in the final accounting, she will be vindicated. She may speak daggers, but uses none. She is associated always with ways of holiness. Her care for Rabbi Elazar throws into relief the Rabbis' mistreatment of him; her ethical behavior highlights their pettiness. Her care for Elazar's body in death is ultimately responsible for the protection of the people of Akhbariya from wild beasts (and demons!), even as her concern enables Elazar, even in death, to continue to render judgment. And it will finally be as a result of her loyalty to Elazar (and her rejection of her suitor, Rabbi Yehuda ha-Nasi) that the Rabbis will have to begin a conversation in which distinctions will be drawn between the character of the scholar and that of the pedant, between the nature of learning and that of wisdom, and between the character of the solitary intellectual and that of the teacher. The behavior of Elazar's wife demands that the Rabbis reconsider their own standards. I am angry with the Rabbis for denying this wonderful woman a name. Clearly they are fearful of her power. And I am concerned that if I give her one now I will simply be continuing the destructive patriarchy the Rabbis represent. Alas, she must remain anonymous, a wonderful woman whose own story has been usurped and retold by men.

The denial of voice is a particularly cruel oppression: the cries of physical repression and pain (as we will soon hear) are readily recognized, but the oppression that derives from the denial of voice remains forever silent. This denial is a particularly powerful form of tyranny practiced regularly in education as we continually battle over the identity of the culture that the schools must transmit. The bitterness of the dispute is reflected in the long and vituperative debates over multiculturalism in the classroom. Indeed, "our" has never reflected the appropriate possessive when referring to the artifacts of the "culture" in which the classroom inhabitants were educated. Our culture has been always "their" culture. When "our culture" is referred to by politicians and pundits and pedagogues, it is usually a white, male, Protestant, Northern European culture to which they refer.

It is certainly true that the story of American education has long denied the female voice. This absence from the discourses has impoverished our schools and obscured our ideas of education and wisdom. For many years, I have sought out the feminist voice in education as an alternative to the authority of the male. I am myself male and have been roundly castigated for the dominance I practiced, sometimes without awareness, though always with privilege. In 1988, I wrote a paper about this topic exploring the means by which my patriarchal past gendered the male self and thereby, gendered the male teacher. This study remains an important and ongoing enterprise in my academic life. I have these yet-unpublished works in my desk drawer alongside a huge pile of uncanceled gender tickets.

I have frequently, and with good reason, suffered for my patriarchy in both public and private spheres and have been castigated for it pointedly sometimes, even in public; I remain troubled at home concerning the places may daughters can assume in the world. I live at present in a household where no one looks like me and against whom bathroom doors now shut; I begin to comprehend the limits of my gendered understandings. And I discover in these stories in Talmud an acknowledgment by the Rabbis, albeit highly coded and circumspect, of the power and knowledge that women may assert were it not to be so controlled. Though far from guiltless in their abuse of power over women's bodies and voices, the Rabbis in Talmud also give evidence of the conflict between their authoritative desire to assert control over the bodies and voices of the women who frighten them and their determination to establish the ethical bases to which they aspired—the demand to become a holy people. It is in these interstices that a movement toward equality can be detected and a more just ethics be asserted. Elazar's wife's existence is eminently ethical.

The tales told of Elazar's wife offers some insight into the Rabbi's conflicts. Elazar's wife is strong-willed and mundane in the best sense of both terms. That is, she is ultimately concerned with physical survival in this world. Finally, the Rabbis will agree with her that without the body the soul is useless, and they will acknowledge that a human being is a body animated by a soul, and not a soul inhabiting a body. Elazar's wife flees her home with her daughter to preserve their bodies; it is her devoted care of Elazar's body that preserves it; it is to that body that litigants appeal for judgment. I suspect that Elazar's wife embodies the physical. Though Elazar is himself plagued with doubts concerning the justness of his actions, issues of philosophical truth, Elazar's wife offers no evidence of doubt regarding her priorities—the care and satisfaction of the physical body. Thus, when Elazar's masochism and self-absorption begin to threaten the physical and related psychological well-being of her family, she must abandon Elazar to his misery; for the continuance of her own and her daughter's physical survival, she must leave his spiritually-wrought torment. And though Elazar continues to taunt her with his survival, it is upon her care of his body that he must finally depend at his death. It is her concern for his body that makes possible Elazar's continued capacity to make judgments and ensure the safety of the town. It is her strong-willedness—given evidence in her quarrel with the townswoman—that finally calls the Rabbis to action: Elazar's wife's behavior has humiliated them long enough.

I cannot but believe that the Rabbis' inclusion of this present story speaks of their enormous respect for her efforts and her character. After all, it is of her care for the husband, whom they themselves have ex-communicated, and of their own dereliction of duty in his burial to which this story speaks. It is the Rabbis who are rendered powerless by the townspeople and, then to protect their own reputations, it is they who must violate the sanctity of the holiest day of the year, Yom Kippur, to steal the body of their colleague Rabbi Elazar, and carry it in secret to his grave. Perhaps, we are meant to understand their actions as a form of repentance on the Day of Atonement, though we must recall that in Jewish law burials are expressly forbidden on Yom Kippur. Incontrovertibly, it is Elazar's wife who is the hero of these stories. I think that the Rabbis' inclusion of this story speaks insistently, though equivocally (she is, after all, a woman and the men seem frightened of and therefore, desirous to control, woman's power, as it is in her embodied) of their respect for her. In these stories, she is the focus of the Rabbis' attention and her behaviors are praiseworthy. Finally, as we shall soon see, it is Elazar's wife, whom Rabbi, or Master, would marry after Elazar's death. This Talmudic story and the next that follows are evidence of the Rabbis' respect for her high ethical character, Elazar's holiness, and their own shame.

There is, in Greek historiography, a parallel, I think, to Elazar's wife: Socrates' spouse, Xanthippe. She is known in history and in literature mostly for her querulousness and shrewishness. Like Elazar's wife, her complaints derive from the sufferings she endures as a result of her husband's behaviors. Elazar had squandered the family resources relieving his own guilt, thereby depriving his own family of sustenance and warmth. Xanthippe, too, suffers at the whims of her husband. Yet she is never portrayed as having any agency: her anguished cries are not for herself or their children, but for Socrates, who, however, has never been gainfully employed. Though his name is forever linked to a pedagogical method for which he, in fact, denies ownership, Socrates defiantly rejects his position as teacher and never appears to take any money for his pedagogical efforts. Xanthippe's penury results from Socrates' refusal to engage in remunerative labor. Elazar's wife, at least, was supported on the monies from her father and when they were spent and Elazar showed no desire to replenish their resources, she was finally impelled to leave his household.

Xanthippe remains the silent sufferer whose silence permits the males to continue to slight her. I. F. Stone offers singular sympathy for Xanthippe in his book, *The Trial of Socrates* (1989). Spoken of badly by Socrates, ignored by Plato, and rendered nameless in the final farewell in the *Phaedo*, Xanthippe is cruelly portrayed as a nag and termagant. Unlike the Rabbis' account of Elazar's wife, Plato displays no conflict in his contempt for Socrates' wife. Her silence is eternal and his condemnation forever. Stone says that even in the dialogues, Plato allows Xanthippe to show sympathy for Socrates, which transcends her own private grief even though no sympathy is ever offered for Xanthippe or Socrates' now fatherless children. Stone laments Xanthippe's maltreatment by both Socrates and Plato. "In the farewell discussions of the Phaedo, the philosopher and his disciples show themselves capable of deep feeling, but only for themselves" (1989:193). Xanthippe is excluded from all participation in Socrates' death or burial; Xanthippe is denied all access to her husband; Xanthippe is denied any connection to Socrates' life or death. I am curious how Socrates' story might have sounded had Xanthippe been allowed to enter the conversation? I wonder, what might have happened to the Socratic method?

Plato's silence regarding Xanthippe can be contrasted with the stories that give insight into the respect due Elazar's wife. In these stories, Elazar's wife will not be silenced, will not be abandoned, and will not be pushed aside. Finally, it is she who assumes all responsibility for her excommunicated husband; it is her loyalty to his body that protects his soul and the well-being of the town. The Rabbis cannot help but express admiration for the ethics of Elazar's wife.

There is to my mind no question that it is appropriate to imagine that the story told here in Talmud is narrated to reflect upon the questionable character of the Rabbis. They are shown, I believe, in all of their pettiness and mean-spiritedness. Their motive to inter Elazar is motivated by concern not for their colleague but for their reputations that might be deleteriously affected if their ill treatment of Elazar's body becomes public knowledge. Clearly, when their own reputations are at stake, these professionals are quick to act according to public opinion. The invective hurled at Elazar's wife by the other woman (how easily the Rabbis pass over to anonymous women their own ill-will and insecurities) is an indication of how much is already publicly known about their colleague's condition and impels the Rabbis to action. Maybe the Rabbis' urge to act stem from their own fear regarding the treatment they might themselves receive from their colleagues upon their own deaths.

Even great scholars, it would seem, are not immune to human failings. We might even read into this story a portrait of the Rabbis' jealousy; after all, in death Elazar is more powerful amongst the people of Akhbariya than are the living Rabbis. There is a Hasidic tale: The story is told that the Rabbi of Ger was asked by one of his students, "What is the meaning of God's asking Cain why his countenance had fallen? How could Cain's countenance not fall since God had not accepted his gift?" And the Rabbi of Ger answered, "God asked Cain: 'Why is thy countenance fallen? Because I did not accept your gift, or because I accepted that of your brother?'"(Buber 1967). Cain's despondency derived not from God's rejection of his gift, but from God's acceptance of that of his brother, Abel. Cain would prefer no one to rise as long as it will not be he himself. Today, we refer to this state as *schadenfreude*, a condition, as Gore Vidal says, which means that "Every time a friend succeeds, I die a little." Elazar's reputation disturbs the Rabbis: his inanimate, departed body appears more powerful than their living minds. For the sake of their reputation, it is time to get rid of the corpse.

Reputation is a powerful spur to action and a strong prod to behavior. In his conspiracy to destroy Othello, Iago uses the loss of reputation to threaten the Moor and spur him to destroy himself. "Who steals my purse steals trash; 'tis something, nothing; / But he that filches from me my good name / Robs me of that which enriches him / And makes me poor indeed" (III, iii, 157–61). Iago's strategy is brilliant: he attacks Othello with something against which the Moor has no power to struggle, a shadow—reputation. This public standing is the shade caused by the interception of light; a shadow is a reflected image produced when a solid body intercepts the light of the sun or other luminary. That is, reputation cannot exist without the light of an Other. Reputation cannot exist except as a substanceless and indistinct reflection. Reputation is not essential to character but extrinsic to and imposed on it. It is reputation, therefore, that is ephemeral—it disappears with the loss of light—the loss of the other. Indeed, Iago might have it reversed: it is reputation that is "something, nothing."

But I am again being a bit equivocal: I admire the reputations of others. I appreciate the good reputations of others and am sometimes envious of their position. Of course, position is not identical to reputation as the behavior of our public figures makes us so well aware. But it is often true that those in high position have reputations: their shadows are readily visible. I would seek my shadow if I only knew where to look. If all depends on local custom, then reputation is a response to a specifically cultural, idiosyncratic local custom.

There is no way to know how to gauge the character of my own reputation. Perhaps it is that I have none, but my own self-respect denies this possibility. I watch for signs but I don't really understand what it is for which I look. I think I would have a good reputation if I could only understand what it might entail. I am searching for a shadow. Clearly, reputation is something over which I have, at best, marginal control. Reputation always derives from without; but contentment, what I dream reputation might deliver, comes from within. The Rabbi of Kalev asked Rabbi Yehudah Zevi to tell him the words of his teacher. This teacher had a reputation for wisdom. Zevi told him that the teacher's words were like manna that enters the body and does not leave. But the Rabbi of Kalev was insistent: please tell me what your teacher told you that I might learn as you have learned. So Rabbi Zevi tore open his shirt and (like the Reverend Dimmesdale!) cried, "Look into my heart! There you will learn what my teacher is "(Buber 1967). My reputation lies within the breast of another. It does not belong to me at all. It is my character and not my reputation that resides in me.

I aspire to an excellent pedagogy. I want to be a good teacher. How can I measure my reputation in this endeavor, by my popularity? Ah, but my courses are required. By the grades I give? Are the grades I record considered a reflection of the teacher or the student? If I can consider this question at all, then of what value are grades ever? Whence my reputation? Does it derive from the general happiness and good nature of my students? How can I take any credit or strength from the characters of those who come to me from a history of which I am not a part, or who go into a future that only marginally contains me? Can I measure my reputation by the achievements of students on standardized tests based in established standards in which I have no voice? If so, then my role as teacher becomes that of technician transferring knowledge from one container to another. Shall I assess my reputation from the success of my students in their chosen professions? And by what means will that be measured? It seems easier to await Godot. Soon after the death of Rabbi Moshe, Rabbi Mendel of Kotzk asked one of his disciples, "What was most important to your teacher?" The disciple thought and then replied: "Whatever he happened to be doing at the moment" (Buber, 1947). The greatness of the teacher was in his present engagements and not in his history or that of his students.

And I want as well to be a scholar. And how will that name be called? By the number of publications I accumulate over the years, or the quality of the ideas contained in them? Who will be their judge? By the sales of my books, or the character of those who purchase them? I wonder now if there might not be some contradiction between these two endeavors of studying and living. I recall Thoreau once complained that there was some paradox between the desire to experience the world and the business of writing about it. We would rather be engaged in life than writing it. Here is a wonderful story (Buber 1967). Rabbi Mendel's students asked him why he didn't write a book. He responded:

> Well, let's say I have written a book. Now who is going to buy it? Our own people will buy it. But when do our people get to read a book, since all through the week they are absorbed in earning their livelihood? They will get to read it on a Sabbath. And when will they get to read it on a Sabbath?

First they have to take the ritual bath, then they must learn and pray, and then comes the Sabbath meal. But after the Sabbath meal is over, they have time to read. Well, suppose one of them stretches out on the sofa, takes the book, and opens it. But he is full and he feels drowsy, so he falls asleep and the book slips to the floor. Now tell me, why should I write a book?

How could Rabbi Mendel's book improve the lives of those people?

Iago plays on Othello's insecurities—the shadow that is cast for others to see—and suggests that Desdemona's infidelities will tarnish the reputation he has previously earned through his military activities. Iago's insinuations suggest that our vulnerability is everywhere and forever. There is no respite from public opinion and no defense against it. Until public opinion is aroused, the scholars are content to let Elazar's body alone; once public disclosure is threatened, the Rabbis are pressed into action. I am reminded of the secrecy that attended the escalation of the Vietnam War, the attempt to avoid disclosure of the Watergate conspiracy, the highly secretive funneling of monies to the contras, and the attempt to withhold from the public the sordidness of the President Clinton's passions. And I remember as well the furious activity—and criminal activities—the disclosures inspired. The status of our reputation over which we have little control troubles our ego. Insecurity, I think, is endemic to the academic profession. Working with and disseminating knowledge is the activity of the educational establishment, but though there is no ultimate end to our efforts, we endeavor as if there were such satisfaction. William Ayres told me that each and every morning academics rise out of bed to the business of producing knowledge and seeking answers. As there is no end to learning or to the availability of possible answers, scholars are always berating themselves for having fallen asleep and, like the hare, approaching the finish line breathless and in last place. Or fearful of being arrested by the king's men.

✡

I believe that the educational discourses to be discovered in Talmud have much to say in both substance and technique concerning the issues and conflicts that swirl about in educational circles and in the daily papers. However, Jewish wisdom has traditionally been excluded and silenced in the West. These Jewish discourses regarding education have been rendered absent from curriculum talk despite the traditional Jewish emphasis on study and Torah. Christianity delegitimized, usurped, and reinscribed the Jewish voice and therefore, forever altered Western pedagogical purpose and practice. Paul replaced the authority of the eminently interpretable text with the mediating authority of Jesus. Whereas the Rabbis decreed that meaning was not in heaven, for Paul, all meaning was situated and derivable from there. Meaning was certainly not on Earth. In Jesus, the text is made flesh: whoever believes in me shall be saved. The concern of the Christian exegetes turned from textual interpretation—the very fabric of Talmud—to the primacy of Jesus's words. It is the personal mediation of Jesus that interprets the text; the Puritans insisted in the Olde Deluder Satan Law of 1647 that children be taught to read but only so that they might read the Bible correctly.

In today's schools, it is yet the teacher who knows and the student who must learn. Reading requires courage because the reader must take responsibility for her reading. But if what I read is understood as separate from me, as belonging to someone else, then do I grow fearful that I cannot negotiate the distance between text and self safely so as to attain the right forms? "What do you think is the best title for this passage?" we ask. "What do you think the author meant here?" I wonder, do I dare to take the risk? Today, in so many of our schools, reading represents separation and loss. Reading in schools requires a denial of self and the acknowledgment of the authority of others in representing the world. Reading, in traditional pedagogies, removes the reader abruptly and often definitively, from what s/he has lived through in reading the word. No wonder our students seem to have "reading problems." Getting it right is what classroom reading requires, separating readers from the production of meaning and ultimately making reading impossible. Reading becomes getting it right. If one feels unsafe and unsure, what is too often referred to as boredom turns insidiously to recalcitrance.

The Platonic-Christian notion of the Real—real concepts and real ideas—and a belief in a mind that could grasp these concepts and ideas through reason, persists in the discourses of education and continues to deny and delegitimize the Jewish voice regarding matters of teaching and learning and wisdom. Socratic teaching or learning is at the heart of most progressive and even some less progressive prescriptions for schools, but not the rabbinic method wonderfully displayed in the Talmud. There, conversation may be defined as reinscriptions and reinterpretations of the Book to meet the kaleidoscopic changes in the Jews' situation—in local custom. The Talmud is the written story of that interpretation, of the making of bylaws, and of adding to the store of Jewish legislation. The Talmud is a way of thinking about how to live in the world based upon an ontology of ethics. It is the story of everyday people struggling to be a holy people. It is the story of Elazar's wife and people like her. The question in Talmud is not what is the Word, but how might the Word be obeyed. There are no singular solutions. The Rabbis debate, but they do not resolve.

However, in our schools, the Greek logos on which Western education is based, premises a right and wrong, a correct and an incorrect knowledge. Our children's papers are filled with red "x" marks and harsh comments like, "Needs improvement." Rabbinic thought prioritizes a multiplicity of meaning and offers not knowledge but ways to it. The language of the schools excludes Rabbinic thought; I have spent my entire life in schools. What language was I speaking?

✡

Rabbi sent [a messenger] to propose [marriage] to his [Elazar's] wife, [and] she sent [back] to him: "Shall a vessel which has been used for holiness be used for secular [purposes]?" There [in Palestine] they say: "Where the master of the house hung his weapons, shall the mean shepherd hang his bag?" He sent [back] to her: "Granted that he was greater than I in Torah, but in good deeds was he greater than I?" She sent [back] to him; "Whether he was greater in Torah I do not know. [But] in deeds, I do know, for he accepted upon himself afflictions."

These stories set into relief how much we owe to Elazar's nameless wife: there, it was her care for his body that assured the possibility of his wise judgment; here, it is her concern for the continued sanctity of her body that leads to her refusal of Rabbi's proposal. There, it was Elazar's self-absorbed suffering that provoked her abandonment; here, it is her respect for his suffering that results in her rejection of Rabbi's marriage proposal. There, it was her anger at Elazar's eagerness to invite sufferings upon himself only when he wasn't studying; here, it is her inability to judge whether Elazar's study has made him the better scholar that will provoke the Rabbis into a far-ranging disquisition concerning the nature of scholarship and learning and teaching.

Elazar's wife rejects Rabbi's marriage proposal because, she says, her husband's suffering is testament to the quality of his deeds. His suffering sanctified him and ennobled his deeds, and his marital relations made her body a vessel for holiness. Rabbi is willing to acknowledge that Elazar may have been smarter, as it were, than he, but Rabbi wonders whether Elazar's deeds could be considered greater than his own as well. Rabbi's implication is that if such were not the case, his offer of marriage ought to be acceptable to Elazar's widow. Rabbi assumes, I suspect, that there is a qualitative distinction between engagement in scholarship and engagement in deeds. That is, both events are eminently measurable and thereby comparable. In American educational circles, it was E. L. Thorndike who said that if something exists, it can be measured. And Rabbi indicates that if Elazar's deeds are not (measurably) superior to his own, then Elazar's wife ought to consider him a suitable suitor. Elazar's wife admits to her inability to quantitatively measure and thereby judge Elazar's respective scholarly achievement, but she is certain (as well she should be!) about the degree of suffering he has willfully accepted upon himself. The degrees of suffering, she suggests, gives testament and character to the quality of the deeds. Since Elazar suffered so greatly, so, too, must his deeds have been great. It is in this case deeds and not scholarship on which Elazar's wife bases the relative merits of the characters of the two Rabbis. Perhaps, she implies, that study in the absence of deeds—with the hint that deeds always produce some suffering—is insufficient.

It is significant to me that the Rabbis would introduce this discrimination between study and deeds through the female voice; it represents to me an implicit self-criticism of the Rabbis who have, perhaps, neglected the muddiness of their own world for the purity of study. It is, in this case, eminently appropriate that it is the woman, Elazar's wife, who brings to the fore this issue because it is she who, during the afflictions of her husband, and now in his absence, continues to maintain the physical life of her family; as it was she who had maintained the body of her deceased husband exactly because of Elazar's (apparently legitimate) fear of the Rabbi's neglect of it. Elazar's wife is herself no stranger to suffering, though she has had little time for study. Her commitment to the material life has kept her from the study hall (and even perhaps the tavern!) and she here expresses her inability to measure the relative (and subtle) degrees of learning that would distinguish Elazar from Rabbi. She has no way to measure the degree of their scholastic accomplishment. However, Elazar's wife's engagement with the physical life has prepared her to recognize Elazar's superiority as a person. After all, he has invited upon himself physical suffering. But as we will see below, it is not the suffering in and of it itself that is praiseworthy, but rather, the sources of that suffering.

There would seem to be a contradiction in Elazar's wife's behavior: earlier she had complained that her husband's willingness to suffer while he was at home was causing herself and daughter bitter hardships; she abandoned Elazar to his sufferings. However, in this story, she seems to praise her husband *because* of his sufferings. On the one hand, she has forsworn Elazar because he suffers and on the other hand she rejects ha-Nasi's proposal because he hasn't suffered enough.

Perhaps, though, there is no contradiction: what had earlier provoked Elazar's wife was her husband's attempt to pass his sufferings off onto her, to relieve his own discomfiture by increasing hers. Indeed, the earlier story recounts that Elazar would willingly put aside his sufferings each morning so as to return to the study of Torah, but each night he would invite his sufferings back upon himself, requiring nursing from his wife and daughter. Perhaps this present story suggests that study without personal suffering is an empty learning, but that a suffering that will not be silently endured is an empty deed. Suffering that must be advertised is inauthentic.

A Hasidic tale: On every Sabbath eve Rabbi Hayyim of Kosov danced before his disciples. His face was aflame and they all knew that every step was informed with sublime meanings and effected sublime things. Once, while he was in the midst of dancing, a heavy bench fell on his foot and he had to pause because of the pain. Later they asked him about it. "It seems to me," he said, "that the pain made itself felt because I interrupted the dance" (Buber 1967). In the course of our lives, we must expect a great deal of contingency and therefore, a great amount of suffering. Even in our celebrations we are subject to accident and even misfortune. But, Rabbi Hayyim of Kosov suggests, as long as we are alive and active in the glorification and repair of the world, then we will not experience the pain. Suffering is most experienced when it is the subject of our consciousness. Another tale: Rabbi Moshe Leib said: "The way in this world is like the edge of a blade. On this side is the netherworld, and on that side is the netherworld, and the way of life lies in between" (Buber, 1967). Our existence is a thin, thin path and the way too often treacherous—but then, consider the vast eternity of the alternative.

It is my experience of late that the response to my query of another, "How are you?" is a deep sigh, a mournful shake of the head and a breathless, "I'm busy!" I am intimidated by this response, and my first inclination is to pack my bags and flee to my father's home. I recognize that we are active, but why are we busy? Of course, like much conversation, "I'm busy," may be meant to ease one into more intimate conversation, but I am concerned in what direction that discourse may be heading at such a beginning. What could I possibly respond to the Other's "I'm busy?" but "I'm not," or "Me, too." A certain tone has been set here that facilitates only the sharing of misery. Whenever I hear this reply, I want to inquire: "What would you posit as an alternative to your "busy-ness?" What does "busy" mean, anyway? What would people be doing if they weren't busy? I am on sabbatical and ought not be busy, but I certainly know that my life is full. When someone complains to me from the oppressiveness of their busy-ness, I wonder if they are suggesting that the amount of their sufferings indicates the superiority of their deeds to my own. Are they suggesting that no one suffers as they do, or that no one is as busy? I think that many people have taken to dressing up in their sufferings for all to admire.

And lately I begin to suspect that the statement "I am busy" is a way of saying "You are not high on my list of priorities. I am focused elsewhere."

✡

"In Torah." What is that?

When Rabban Shimon ben Gamliel and Rabbi Yehoshua ben Korhah sat on benches, Rabbi Elazar the son of Rabbi Shimon and Rabbi sat before them on the ground. They [the sons] raised objections and answered [them]. They [the sons] said, "We drink their water yet sit on the ground!" They [the Rabbis] made benches for them [and] raised them up. Rabban Shimon ben Gamliel said to them: I have one dove among you, and you wish to deprive me of it!" They demoted Rabbi.

Rabbi Yehoshua ben Korhah said to them: "Will he who has a father live, and will he who has no father die?" They also demoted Rabbi Elazar the son of Rabbi Shimon. He was offended, [and] he said: "Do you consider him my equal?" Until that day, when Rabbi said something, Rabbi Elazar the son of Rabbi Shimon would support him. From then onward, when Rabbi would say, "I have an objection," Rabbi Elazar the son of Rabbi Shimon would say to him: "Such-and-such is your objection, [and] this is your answer. Now you have encircled us with bundles of arguments that have no substance." Rabbi was offended, [and] he went [and] told his father. He said to him, "My son, do not let it distress you, for he is a lion the son of a lion, and you are a lion the son of a fox."

There are two wonderful and inseparable issues here. In the previous story, Elazar's wife admitted that she did not know if her husband was greater in Torah than Rabbi, though certainly he was greater in deeds than he. What is the relationship between study and deeds? The issue of what does "in Torah" mean is now raised, though I do not think it is resolved here. Indeed, what does it mean to have "greater" knowledge of a text—of THE text, in fact? Interestingly enough, in the present story the issue appears to be abandoned as soon as it is raised; I do not think the Rabbis have the slightest idea how to quantitatively measure such achievement. I am certain that standardized tests had yet been developed. Instead, to the question, what does "in Torah" mean, the Talmud responds with a story concerning the two Rabbis, Yehudah and Elazar, sons of their illustrious fathers. This raises the second issue: the politics of rank. Does being greater in Torah have anything to do with the respective ranks of the scholars? Can we tell anything about a person's knowledge from his/her position in the hierarchy?

The arrangement of study presented here in this story is simple: the eldest and most respected scholars sit on chairs, and the students sit on the ground at the feet of the Rabbis. This arrangement calls to mind so many of the classrooms in which I have taught and in which I have observed: students sitting quietly—listening, absorbing, even sometimes enthralled, but saying little—at the symbolic feet of the scholar. How do they know they are in the presence of a scholar? Of scholarship? Is it really their placement in chairs that makes the scholar? The question arises: how does one attain finally to the chair! In our professional schools today, one acquires the chair by successfully passing a number of courses and standardized tests, even writing a dissertation. These accomplishments attest to the possession of at least a modicum of wisdom. But from their position at the scholar's feet, Elazar and Rabbi raise objections about what they say to those in the chairs. That is, the students engage in disputation with their teachers and provoke conversation. These two sons challenge their fathers, both their physical and intellectual progenitors. It is, ironically, not what the sons know, but what they acknowledge they do not know that earns these young scholars an invitation to their seat on the bench. But this promotion makes them subject to the Evil Eye—the critical gaze, I suppose, of those yet on the ground.

Elazar and Rabbi have been schooled by their illustrious fathers; the learning of the fathers has passed to the sons. However, here the Talmud refers not to the influence of heredity but to that of environment. It is not mere physical propinquity to the fathers that leads to the sons' intellectual powers, but to the literate environment, which the fathers maintain, to which they assign value, and in which they have raised their progeny. The significance of environment on the growth of our children (the nurture) is common knowledge in today's educational circles: children surrounded with books and learning, succeed better in school than children for whom this advantage is absent. Sitting at the feet of these two great scholars, Shimon ben Gamliel and Rabbi Shimon ben Yohai, the son's agility and perspicacity of minds becomes clearly evident. In their discourse with their father-teachers and their challenge to their teacher-fathers' tenets, the sons not only sharpen their own minds, but also serve as stimulus to their elders' continued learning. This is not unlike the story told above of Resh Lakish and Johanan: good students are good teachers for their teachers. The best student challenges the teacher. A good teacher is first a good learner. The moral of this story for us is simple: there must be places provided for students to engage in equal discourse with their teachers about issues of substance and meaningfulness.

In the situation presently under discussion, the brilliance of the sons' disquisitions causes the other Rabbis to insist that they be risen in rank and sit on chairs equal to their fathers. Rank is premised on ability. But one of the fathers, Shimon ben Gamliel, voices an objection: elevating his son, Rabbi, in rank makes him vulnerable to the gaze of the Evil Eye and imperils the son's safety. And so, at his father's request, Rabbi, the son, is demoted. Then the issue is raised: since Elazar's father, Shimon ben Yohai had since died and could no longer speak for his son, Elazar still seated on the bench, is vulnerable. And so Rabbi Yehoshua assumes responsibility for Elazar's welfare and insists that Elazar, too, be demoted so that the Evil Eye not detect him. Both sons return to their place on the ground at the feet of the scholars.

I have spoken above about the Evil Eye. There, because Rabbi Yohanan gazes at women exiting the ritual bath, the mikvah, he is susceptible to the Evil Eye. But, Yohanan objects, since he is a descendant of Joseph, he is therefore immune to the powers of the Evil Eye. Since teachers are always the object of the gaze and are forever gazing, a vulnerability to the Evil Eye would render them incapable of functioning. We acknowledge, as well, that the gaze of the teachers must be unsullied by prurience. Teachers must be invulnerable to the Evil Eye. But here in this story however, exposure to the Evil Eye occurs from a promotion in rank: clearly the Evil Eye notices a new voice that challenges the hierarchical order. The Evil Eye takes advantage of this increased visibility. Rank is, after all, a sign of order, and order, after all, must be maintained. Elazar himself had discovered thieves by discovering them in the tavern when they should have been elsewhere. In the hostile environment of the academy, these scholars are endangered without the protection of their masters— their fathers. To be caught in the gaze unawares invites danger. I have myself often felt threatened by the eyes of my students and even colleagues. Where would they have me be: on the chair or on the floor? Where would they prefer to be: on the floor or raised to the chair? Teach me, they say. Sit in the chair, they say, and tell me the truth! I stand in front of the classroom and during the semester of sixteen weeks must deliver a lifetime of learning and thought. When I inevitably fail, I am subject to ridicule and scorn. And when I tell them only what I know, I am dismissed as inadequate. Their gaze endangers the work of the classroom, which is our learning. The Rabbi of Kotzk said: "Everything in the world can be imitated except truth. For truth that is imitated is no longer truth" (Buber, 1967). But if I demand that students pursue in their work their own truths, they tremble in fear and anger having been led off the curricular path. They would have the answer before the question is asked. I would have them raise objections, and they object to the request.

Finally, in this story, the promotion does not enhance the reputation of the sons but rather, forces them into the gaze of others. It is possible that in becoming a teacher, by rising, as it were, into the chair, learning is threatened. Wouldn't the gaze of the Other be as dangerous as the Evil Eye? What is expected from those in the chairs?

✡

To be better "in Torah!" What could that mean? Once again the Rabbis model a negative example: the promotions and demotions of Elazar and Rabbi end in dissension and meaningless debate. Vying for intellectual advantage and superiority, the Rabbis descend into petty bickering and scholastic pettiness. They participate in esoteric and effete disquisitions; they debate for the sake of debate. "Now you have encircled us with bundles of arguments that have no substance." The scholars are entrapped in barren argument. Thus it is that the issue, what does "In Torah" mean, is not abandoned so much as it is elaborated by the negative. Having been demoted, Rabbi resorts to esoteric and abstruse argument. Competing for academic brilliance, Rabbi produces only polemic. Talmud responds that "In Torah" does not refer to rank or to quantity of knowledge nor to the ability to produce abstruse argument. A scholar must know more than how to construct esoteric polemic. It is not the position in the chair that makes the scholar; indeed, sometimes the position in that chair endangers learning.

Nonetheless, because Shimon ben Yohai was such a great teacher, Talmud suggests that Elazar must be a better student than Rabbi, though we have not yet determined what work a great teacher does, because we have not understood yet what "in Torah" means.

Perhaps it is possible to view this story as an elaboration of the previous one; once again the pettiness of the Rabbis is evident. What they seem inordinately concerned with is status: who should be sitting on the chairs and who on the ground? The equation of rank and learning is a powerful one in the culture of the school as well; promotion is a consummation devoutly to be wished. But these stories call into question the very ranks they establish. I have previously discussed a story that contrasts the Greek portrayal of Xanthippe and the rabbinic characterization of Elazar's wife. I would like to offer another contrast between the Greek and Hebrew perspectives, this time concerning the role of the teacher. The Talmudic version appears above: it first narrates the history of Resh Lakish and his brother-in-law, Rabbi Johanan. In Johanan's inability to survive without Lakish, we understand the intimate relationship between the teacher and the student. Without a student, one cannot be a teacher. In this relationship, the Talmud defines learning as a result of intimate intellectual interaction, and Talmud portrays education as an enterprise in which all participants have a *share*. For the Rabbis of the Talmud, education is an activity in which benefit arises not from the satisfaction of an interest but from commitment with and to another in the process of learning. The benefit of learning accrues not from the obtaining of an object by one or the other, but from the engagement *in* the activity *with* another. It is the sharing from which education occurs and not from the acquisition of the product of that sharing. In the story concerning the promotions and demotions of Elazar and Rabbi, it is the students' challenge to the scholars that earns them ascension to the chair.

But the Socratic (Platonic) tradition in which I have been educated argues a different connection between teacher and student and therefore, assumes a different concept of learning. Socrates (and Plato) suggests that we have lost all of our knowledge and we must spend our lives retrieving it. Socratic pedagogy is based on the idea that education pursues the true and the good that absolutely exists and may be learned. For the Greeks, education is a matter of accumulation; it accrues to the individual engaged in it in a sort of quantitative measure. Education is a private enterprise and must be privately accomplished. Socrates, indeed, disowns his own role as a teacher and disavows any share in the educational endeavor. Socrates argues that if learning will occur, it is the responsibility of the learner to accomplish it. He, for one, will not be held responsible. He cautions: "If any one of these people becomes a good citizen or a bad one, I cannot fairly be held responsible, since I have ever promised or imparted any teaching to anybody, and if anyone asserts that he has ever learned or heard from me privately anything which was not open to everyone else, you may be quite sure that he is not telling the truth." Here, truth is available to the seeker as a product to be owned, so to speak; one has only to look properly for it. And whatever might be lost and/or gained by the other is of little interest to Socrates; disavowing his role of teacher, he denies the reality of the student.

But for Yohanan and Lakish, education had been a relationship and the acquisition of knowledge as participation. So, too, in this story of Elazar and Rabbi; they are students because they can be teachers.

✡

> And this is what Rabbi said: There are three humble people, and these are
> they: My father, and the sons of Betera, and Jonathan, the son of Saul.
> "Rabbi Shimon ben Gamliel"—this [refers] to what we said [above].
> "The sons of Betera"—as the master said: "They placed him [Hillel] at
> the head and appointed him Nasi over them." "Jonathan the son of
> Saul"—for he said to David: "And you will be King over Israel, and I will
> be second to you."
>
> From what? Perhaps Jonathan the son of Saul [said this] because he saw
> that the people were drawn after David. [And perhaps] the sons of Betera
> also [acted as they did] because they saw that Hillel was superior to them.
> But Rabban Shimon ben Gamliel was certainly humble."

I think the Talmud has returned, albeit obliquely, to the issue of what
does "in Torah" mean. Rabbi's father suggests to him that he should not
be jealous of Elazar's intellectual prowess because Elazar is the son of a
lion whereas he, Rabbi, is merely the son of a fox—clever, but not as
intellectually formidable and brilliant as his friend. Rabbi refers to the
humility that his great father surely possessed; humility is a trait intrinsic
to the scholar. Perhaps that is the trouble the Rabbis always have with
Elazar—though he is admittedly brilliant, he refuses to be humble. It is
his hubris that disturbs them rather than his ability. Of course, Elazar is
not alone in his pride: Rabbi acknowledges only three examples of true
humility in all of history. Humility, the story suggests, is a characteristic
of the great scholar. And Rabbi defines humility with three telling examples: the first is his father, Shimon ben Gamliel, who acknowledged only
to his son and in the privacy of their home that Shimon ben Yohai was
the greater scholar. The second example of humility was Jonathan, who,
though the son of King Saul, acknowledged that his friend, David, would
be king. Finally, the sons of Betera are exemplars of humility, because
though a family of famous sages, they willfully abdicated their leadership
position to Hillel whom they considered more brilliant. Humility seems
to be defined as the acknowledgment that one does not have the answer;
humility is the acknowledgment of an Other whose answers satisfy us.
Humility is admitting our apprehensions to ourselves and not boasting of
them as the achievement of humility.

There is a cliché in academic circles that teachers must learn to say "I
don't know." But in fact, our pedagogies—and especially with the institution of organized standards—are designed to prevent the difficult questions from ever being asked in the first place. I think the Rabbis would

aver that the willingness to admit that we don't have answers to some questions does not condone our pride in assuming answers to so many others. Humility is the acknowledgment of how much we do not know. It is told of Rabbi Elimelekh that he once said: "I am certain to have a share in the coming world. When I stand in the court of justice above and they ask me: 'Have you studied all you should?' I shall answer, 'No.' Then they will ask: 'Have you prayed all you should?' And again I shall answer, 'No.' And they will put a third question to me: 'Have you done all the good you should?' And this time, too, I shall have to give the same answer. Then they will pronounce the verdict: 'You told the truth. For the sake of truth, you deserve a share in the coming world'" (Buber, 1967).

I do not think that we need think of the "coming world" as that which exists after death. Tomorrow is always the coming world, and our humility always reminds us that we have not yet done enough to earn it; our honesty, which is often our humility, however, carries us to the world to come—to tomorrow—in our effort to keep on keeping on.

And so the objections to Rabbi's examples of humility are telling; we are concerned with a recognition of humility. It is said that neither Jonathan nor the sons of Betera were truly humble: all had objective reasons for abdicating their claim to higher position. In each case, popular opinion was clearly in support of their rivals: an advocacy for the national leadership of David and a public awareness of the intellectual authority of Hillel. The decision made by Jonathan and the sons of Betera was a politically astute one and not an example of simple humility. However, it is claimed, since Shimon was speaking only to his son, it was not obvious that anyone else could hear. Therefore, since none knew of his humility, it was accounted the greatest. Even humility raises one's reputation as long as that humility is spoken of by another!

✡

Rabbi said: "Sufferings are precious." He accepted upon himself thirteen years [of suffering], six of kidney stones and seven of scurvy. And some say: Seven of kidney stones, and six of scurvy. The stable master in the house of Rabbi was wealthier than King Shapur. When he used to place fodder for the animals, their voices would carry for three miles. He would aim to place [the fodder] at the same time that Rabbi went to the toilet, but even so his cry would rise over their cries, and the seafarers would hear them.

But even so, the sufferings of Rabbi Elazar the son of Rabbi Shimon were superior to those of Rabbi. For [those] of Rabbi Elazar the son of Rabbi Shimon came through love and went away through love. Those of Rabbi came because of an incident and went away because of an incident.

"They came because of an incident." What is [meant by] this? For a calf which they were bringing to slaughter went [and] hung its head in the corner of Rabbi's garment and cried. He said to it: "Go, for this you were created." They said: "Since he shows no compassion, let sufferings come upon him."

I am intrigued by the Rabbis' focus on suffering as a bellwether of virtue. I have myself contradictory feelings concerning suffering. On the one hand I despise it. I would wish it away. Once, when Rabbi Yonatan was asked whether suffering was dear to him, he responded, "Neither it nor its reward." I concur. On the other hand, it is sometimes quite satisfying (even comforting) to issue howls and laments that rise over the lowings and bellowings of my fellow creatures; it indulges me to bemoan the troubles I've seen. Often, we actually define ourselves by our suffering. In conversation we tell tales of our miseries and measure our reputations by the plenitude of our sufferings. It is a perverse game of one-upmanship; on the one hand, the one with the alleged greater suffering is the winner. On the other hand, it is an exquisite pain.

However, I cannot help but smile at this depiction of the content of Rabbi's sufferings. I believe there must be in the world a better way to portray his pains than by the graphic description of his excruciatingly painful excretory dysfunctioning. It is a much more wonderful John Dewey I know who at ninety-two years old plays a bit painfully on the floor with his child, or who in public bemoans his lack of influence, than the John Dewey who suffers noisily from painful digestive problems or agonizing flatulence. Rabbi's suffering must be extreme, but the Talmud undercuts his anguish in the almost comic depiction of his cries, which surpass even the noises of the beasts and which can be heard even by those at sea.

It is not the quantitative measure of the degrees of sufferings by which it may be adjudged, however. Rather, the Rabbis measure the quality of the suffering of Elazar and Rabbi by the manner in which each bears his sufferings and by the psychological origins from which that suffering arose. The nature of their sufferings, after all, has something to do with the character of Elazar and Rabbi as scholars: Elazar's sufferings are superior (in quality) to those of Rabbi; Elazar may be the greater scholar as a result. Recall that in their discussion, the Rabbis are yet engaged in an elaboration of what it might mean to be superior "in Torah," and as the earlier story shows, humility is one characteristic. Suffering, I suspect, may have humility as one component, but all suffering is not equal; for some, suffering, may contain not a little hubris. Elazar's sufferings, we are told, derived from love and went away out of love. Despite the assurances that the executed man deserved his fate, Elazar invited his sufferings upon himself because of his complicity in the man's death. Elazar's suffering derives from his unsolicited concern for the Other; Elazar pursues his sufferings willingly because another suffers. Furthermore, Elazar's suffering is experienced at night when only his wife and daughter can witness his pain; his is not a public display of travail.[1] Elazar's sufferings are banished every morning because of Elazar's love of the study of Torah.

But Rabbi's sufferings, we are told, resulted from his failure to love, and Rabbi howls his sufferings into the world. Rabbi, unlike Elazar, does not keep his sufferings within the confines of his private domicile; rather, Rabbi bellows his pain even over the lowings of hungry animals.

In these stories, suffering is a consummation to be valued most devoutly when that suffering is a) self-inflicted; b) borne out of love; and c) endured in (relative) silence. The experience of suffering does not in itself inspire admiration or sympathy and need not be indicative of a virtuous soul. In and of itself suffering has no meaning. I think this reminds me that an intention without an action is meaningless and that a good deed accomplished with bad intention remains yet a good deed. And a bad deed accomplished with good intention is yet a bad deed. Maimonides argues that the highest form of charity is that given without any expectation of recognition or reward. Suffering that is broadcast serves only the sufferer and may be not a sign of love and humility, but rather, a sign of pride and self-aggrandizement.

Rabbi had held that sufferings are precious; he believed that the failure of Elazar's body to decompose even after death resulted from his willingness to suffer. Rabbi here seems to accept that suffering of itself is a reward, but that was before he experienced suffering. Rabbi's statement, I think, represents a naïve idealism; he spoke without the experience or personal knowledge of suffering. Indeed, Rabbi lives in fabulous ease: it is said that Rabbi's wealth is such that even his stable master became rich. Yet despite his learning and his wealth—or perhaps because of these accomplishments—he has lost sympathy with the smaller participants and events in life: he callously dismisses the sufferings of the calf. Those in Heaven insist that he suffer so that humility might be learned. Humility is a great equalizer.

Rabbi's sufferings proved not as precious as he had once imagined: he endures horrible and humiliating pains. Nor was Rabbi's reputation enhanced by his suffering. This story suggests that Rabbi's sufferings *become* precious because they led to the very behavior with which it is said that Elazar started—love and compassion.

I am interested in the learning the scholar derives from suffering. I know not a few scholars whose wails can be heard by sailors at sea. I myself often compete with the lowings of animals. Perhaps the Rabbis recognize here that suffering remains a regular experience of life, but that the quality of suffering defines the sufferer. And the quality of suffering gains its character from its cause, which then in turn, inspires a response by the one who suffers. That response, as exemplified by Elazar and Rabbi, ought to result in a further engagement in learning, compassion, and justice. The scholar, of whom I may be one, must suffer: we recognize the world for what it is, and like Abraham we search for ten righteous people—fully aware that we may not be one of them. We acknowledge our share in Sodom and suffer for it. Our suffering must derive from love and not from detachment. It must be borne silently and banished in the morning so that we may continue our work in the world. And that work ought to be the world's repair.

This portrait of Rabbi depicts him in almost comic relief; he is portrayed as callous and unfeeling; his suffering borders on caricature. Those seafarers who overhear his painful cries recall to me those sailors who brought succor to Elazar. Here, however, they can bear no balm for Rabbi's sufferings; they remain merely passive bystanders to his suffering. He must become something new before they can bring any relief for his pain. Rabbi's sufferings are depicted as qualitatively different in character than those of Elazar; those of the former derive from his disdain for the world and those of the latter from his engagement with it. I think they may even suffer in a similar painful, physical manner; there may not be a quantitative distinction between their sufferings, but there are definitely qualitative differences. Rabbi Elazar's sufferings arose and went out of love; Rabbi's sufferings came out of contempt.

For over forty years I have been enamored of one particular folk song. I heard it first from the Chad Mitchell Trio. It is a Yiddish folk song entitled "Dona, Dona." In studying this piece of Talmud, I have learned that the song derives from this story of Rabbi Yohanan. The changes are significant; the verses of the song go, in part:

> On a wagon bound for market
> There's a calf with a mournful eye.
> High above him there's a swallow
> Winging swiftly through the sky. . . .
> "Stop complaining," said the farmer
> "Who told you a calf to be?
> Why don't you have wings to fly with

Like the swallow so proud and free?". . .
Calves are easily bound and slaughtered
Never knowing the reason why.
But whoever treasures freedom,
Like the swallow must learn to fly. . . .

Rabbi cold-heartedly dismissed the suffering of the calf and is himself afflicted with sufferings as a result. But whereas Rabbi callously abandoned the calf to its fate—to its sufferings—the song acknowledges the suffering but urges resistance to it. Suffering is not to be simply borne; it must be resisted. I think the Rabbis have suggested something like this as well: suffering, they aver, is to be used and not merely accepted. In this way, suffering may be opposed. Employing language that echoes the Book of Job (stop complaining!), the farmer taunts the calf with his biology. Like Rabbi, the farmer condemns the calf to its fate. But the lyricist speaks in the third verse of overcoming biology with resistance and struggle. For Elazar's wife, that resistance meant leaving home rather than suffer passively from her husband's solipsism; Rabbi Yose fled to Laodicea to avoid heinous employment by the government. For the partisans in the Warsaw Ghetto, resistance meant armed struggle. In the song, resistance is learning to fly like the swallow.

For Elazar, suffering is overcome by study. I am a teacher. I've seen troubles all my days. This world is no Eden: however, I promise students that learning is the movement away from suffering and towards freedom; I promise learning as the experience of freedom. My students moan that they are suffering. I do not think learning will relieve sufferings—even swallows experience the hardness of life—but they do occasionally soar. Perhaps in the thrall of study, sufferings can be dismissed. I wonder how do I teach this? But sufferings must also be invited back at some time and regularly so. I wonder how do I teach this? Our sufferings must derive from our burdens and must not burden others. I wonder how do I teach this? Suffering reminds us of our humanity; suffering makes us human. Rabbi Moshe Leib said, ". . . if someone comes to you and asks your help, you shall not turn him off with pious words, saying, 'Have faith and take your troubles to God!' You shall act as if there were no God, as if there were only one person in all the world who could help this man—only yourself" (Buber, 1967). The Rabbis tell us that to save one person is to save the world. Rabbi denied his own humanity in his cold response to the calf. How could he be a wise person, a tzaddik, and not understand his humanity?

✡

> And they went away because of an incident. One day Rabbi's maid was sweeping the house. Some young rats were scattered there and she swept them up. He said to her: "Let them go. It is written: 'And his mercies are over all His works.'" They said: "Since he shows compassion, we will have compassion on him."

I wonder if it is that suffering leads to knowledge, or is suffering a consequence of some physical condition and has no rewards? Is suffering a spur to, or a result of, behavior? What, if any, are the pedagogical uses of suffering? Elazar had invited his sufferings upon himself as a consequence of his guilt, but every morning he would abandon his sufferings to engage in study; to my mind, the relationship between his sufferings and his studying is oblique. Elazar wills his suffering even as he wills its dismissal. Rabbi's sufferings, however, came because of an incident, and they went away because of an incident. Rabbi's sufferings were imposed and lifted without his accession or responsibility. Did he learn from his suffering and therefore, change his behavior? Is suffering a cause or a result? As the former, suffering is a pedagogical device with immediate and material reward; advertisers already use this ploy regularly. "No pain, no gain," say the Nike advertisements. Educators, too, have long held this opinion. Suffering has long seemed intrinsic to education; the nineteenth century, for example, advocated rigorous disciplining of the mind and character to raise the human to higher ground, as it were. William Torrey Harris, for one, an incredibly influential nineteenth-century educator whose influence is still felt in our modern schools, held that education was not the facilitation of the natural development of the powers of mind along the lines of its natural growth, but the process by which the powers of that mind are altered from their natural character in conformity to the demands of society. The purpose of education was to alienate the person from him/herself so that each might become something else more noble. School subjects were offered not for their practical use or personal interest, but for their ability to develop strict mental discipline. Suffering was intrinsic to the pedagogical process. Our children are yet told, "I know this material is difficult and you are frustrated and cannot immediately see the importance of this subject, I know this is painful, but believe me, trust me, this is good for you." Our children sit for hours every day for almost twelve years in uncomfortable and difficult furniture, studying topics out of context. Emma's homework consumes hours of her time.

And if the latter, and if suffering results from behavior, then suffering is characterized as a byproduct of ignorance. Rabbi had much to learn:

God's mercy extends even to the little rats on his polished floor. We suffer because we presume to know what in fact we do not know. Learning promises to eliminate suffering. We cease to suffer when we learn.

I think either view oversimplifies the world to our children. On the one hand, we suggest that suffering has a completion—it will end when we "mature," or when we finish learning, or when we leave school, but in fact, that is not our experience of the world. There is no end to our learning, and there is no cessation of our suffering. Our suffering does not cease when we graduate formal schooling; rather it merely changes register. If we continue to be life-long learners and to advocate life-long learning and if we acknowledge the continuance of suffering as a spur to this learning, then we ought to have greater sympathy for the travails our children experience in school who work there mostly under compulsion. Perhaps our misrepresentation of suffering accounts for our children's mistrust of us: on the one hand, we romanticize their ignorant innocence, which requires suffering to foreclose it, and yet we promise them a return to that innocence in the purity of an education—an end to their suffering. But, in fact, the reality of our lives displays none of this possibility. If their innocence had led to suffering, then a return to innocence will produce more suffering. On the other hand, we suggest that the development of knowledge eliminates suffering, and yet the continued presence in the world of suffering belies this ideology. I think we are as conflicted about the relationship between suffering and behavior as are the Rabbis of the Talmud.

I do not think the Rabbis have an easy answer to the value of suffering. It would seem that for them suffering is inevitable—perhaps it the experience of suffering that makes us endeavor as humans. Perhaps it is our knowledge of suffering that makes us human. Adam and Eve in the Garden had no motive to strive and to seek. As far as they knew, all was perfect and all was true; all that they required had been provided, and all was at their command. But once they had encountered evil, then that evil could not be forgotten, and the garden no longer could serve as an isolable environment. That is, if Adam and Eve brought evil *into the garden*, then there could be no escape from evil. From within the Garden they would have to confront themselves and evil. And if there were evil in the world, then they must also go out and struggle against it. The only way to confront the evil in the world was to meet it. Adam and Eve had to leave the Garden for their own benefit and for the care of the garden. Indeed, without human effort the garden would become overrun with evil; without human effort there would be no garden. Adam and Eve accepted suffering upon themselves. Perhaps it is not so simple to say that Adam and

Eve's sufferings came because of an incident: the eating of the fruit of the Tree of Knowledge. It was not disobedience that led to suffering, but to an engagement in the world. But what else could they have done?

Did their suffering lead to learning? Genesis does not explicitly say, and so we must interpret. Certainly, Adam and Eve continued to suffer— we need look no further than the grief they experienced from the fratricide within their family. I learn from Adin Steinsaltz (1992) that Adam and Eve were sent out to work the earth—to dig, to weed, and to draw forth out of the earth. They had to sort out the thorns and thistles and weeds in order that the good and the sustaining might grow. But Adam and Even had to first learn the difference: as they did not know that they were naked, neither did they know the noxious from the good, the deadly weed from the sustaining plant. *Their sufferings came because of an incident*, but it was not the eating of fruit that led to their suffering. Rather, it was because they had so much to learn in the world that they suffered. In suffering, they learned, though their products—Cain and Abel, for example—did not always reflect the learning of their parents. The freedom, which is responsibility, that Adam and Eve acquired in their act of eating of the Tree of Knowledge, must be (l)earned by each individual. Each must suffer by him and herself. Suffering may, but need not, lead to learning. As this is a material world, sufferings come because of an incident and as this is a material world, so, too, do they leave because of an incident. The process is cyclical and ever changing.

I do not know that there is any value to suffering, but perhaps it can be made valuable.

What is the relationship between suffering and learning? In the case of Rabbi, the connection between learning and suffering is not clear. Rabbi suffered because he had shown callousness to a creature in pain. Suffering is visited upon him, and he lives in torment. Whether he learns from his suffering or as a result of his suffering is not, however, explicitly stated. There is nothing in the second story regarding his compassion for the captured rats which gives evidence that his experience of suffering, imposed because of his coldness toward the calf, in and of itself has led to greater compassion. Indeed, I understand that Rabbi's sympathy in this later story derives not from his experience of suffering but from his engagement in study: as justification for sparing the rats, Rabbi quotes Psalm 145, one of the most popular psalms of the Rabbis. It is included in each of the three daily prayer services. Psalm 145 speaks of God's compassion, grace, and mercy. Since Leviticus commands us to be holy because God is holy, then it is incumbent upon Rabbi to show mercy to all creatures because God is merciful. Rabbi's compassion for the rats is not connected to the circumstances concerning his callousness towards the calf; rather, it is connected to the effort of his study. For Rabbi, suffering is yet suffering, but precious it is not. But it may also be with regard to Rabbi that suffering may be understood as intrinsic to living: as God suffers for God's people, so too must humans suffer for themselves and for God. After all, the world is not perfect and even God has erred at times. Therefore, perhaps Rabbi's sufferings are precious because they make him human. A Hasidic tale: Rabbi Elimelkh of Lizhensk declared that fasting was no longer acceptable behavior. His student objected: the Baal Shem Tov, the founder of Hasidism, fasted often! "Indeed," said Rabbi Elimelekh, "the Baal Shem Tov would leave for a week's study with six loaves of bread and a pitcher of water. He would spend the week in study and in isolation, but when he would return home for Shabbat he would pick up his bag and note its weight. When he looked inside he found all the loaves of bread yet with it. Fasting such as this is allowed" (Buber, 1947). Immersed in good works, immersed in study, suffering may be precious, but it is suffering nonetheless. But in the thrall of study, sometimes the pain is less.

When my daughter sits for hours with her worksheets, how is her suffering precious?

✡

All the years of Rabbi Elazar's sufferings, no man died before his time. All the years of Rabbi's sufferings, the world was not in need of rain. For Rabbah the son of Rab Shela said: "The day of rain is as harsh as the Day of Judgment." And Amemar said if it were not necessary for the world, the Rabbis would have prayed and abolished it. Even so when they would uproot a radish from a vegetable patch, the hole would remain full of water.

Nevertheless, the Talmud tells us, the sufferings of Elazar and Rabbi worked great miracles upon the earth. No man died before his time, and there was no drought, though there was, perhaps, at times, an excess of rain. I wonder though, who would know when it is the proper time for a man to die? The Talmud does not say that no one died, but that no one died *before his time*. The Rabbis could not refer here to an absence of death, or even of suffering. Perhaps what the Talmud refers to is the ability of the scholar to participate in the sufferings of the world and thereby, give succor to the populace. As long as the scholar suffers with the people, then the world appears to operate with some harmony and order. No one died *before his time!* Perhaps to be great "in Torah"—the issue presently central to the conversation—is to share in the sufferings of others. Indeed, not much in the world changed as a result of Elazar's and Rabbi's sufferings; men still died and the earth still oozed with too much moisture. The world did not suffer less, but the people were assuaged in their sufferings. People continued to die and it still rained, but death was not so fearsome nor rain so dangerous. The scholar shares the people's suffering by accepting suffering upon herself; the scholar serves not by martyrdom, but by compassion. Whenever the rabbi of Sasov saw anyone's suffering either of spirit or of body, he shared it so earnestly that the other's suffering became his own. Once, someone expressed his astonishment at this capacity to share in another's troubles. "What do you mean 'share?'" said the Rabbi. "It is my own sorrow; how can I help but suffer it?" (Buber 1947). If it is the other to whom we owe our existence, then how can the sufferings of the other not be our own? It is this that the scholar knows? When God asked Solomon what he desired most in his rule, Solomon prayed for wisdom (1 Kings 3:9). "May you grant Your servant an understanding heart, to judge Your people, to distinguish between good and evil; for who can judge this formidable people of Yours?" But Solomon also knew that "He that increases knowledge, increases sorrow." From this the Hasidim say that a man should increase his knowledge even though it will increase his sorrow.

These stories in Talmud suggest that increasing knowledge does not at all bring comfort; rather, in bearing his burdens, the scholar does not feel their weight. Perhaps that is what is meant when the Rabbis say that Elazar was able to put aside his sufferings to go to study; perhaps in study, sufferings may be shared. Rabbi, on the other hand, because his sufferings came because of an incident—his callous response to the sufferings of the calf—could not put aside his sufferings because by condemning the calf, he had also condemned himself. The scholar must always speak with the people. Rabbi David Moshe explained that God said to Moses (Exodus 19:9), "Lo, I come unto thee in a thick cloud that the people may hear when I speak with thee," because God required Moses to remain in direct contact with the people to ensure that the people hear exactly what Moses heard. It is not good, says David Moshe, that the spirit of the tzaddik (the wise one) may mount too high and lose touch with his generation. That is why God masses the dark cloud of sorrow over the tzaddik and sets limits to his soul. But when sorrow—it is a dark cloud and not a bright one in which God's presence resides—descends upon the tzaddik, he finds God even in the sorrow, as it is written (Exodus 20:18): ". . . but Moses drew near unto the thick darkness where God was" (Buber 1967). The story reminds me that the scholar must stay close to the people though this nearness results in sorrow. The scholar suffers and nothing changes; the scholar suffers and the world is relieved. The scholar suffers, but finds God in the suffering. What other value is there in scholarship? Scholarship links the wise to the people and to God and to herself. Scholarship forces us into the world rather than relieves us of it. Indeed, the allure of scholarship endangers the very life of the tzaddik. Without deeds in the world, the wise is without purpose. Once, Rabbi Bunam (Buber 1967) honored a man in the shul by asking him to blow the shofar on Rosh Hashanah. The man began to make lengthy preparations to ensure that each note he produced was meaningful. After some few minutes the Rabbi cried out: "Fool, go ahead and blow!" Better to make a small and imperfect sound than prepare forever to achieve the perfect sound and say nothing.

The students sit quietly and afraid. They are afraid to make an imperfect sound. They would only make the faultless timbre. I want to shout, "Go ahead and blow!"

✡

Rabbi happened to come to the place of Rabbi Elazar the son of Rabbi Shimon, [and] said to them: "Does that righteous man have a son?" They said to him: "He has a son, and every prostitute who is hired for two hires him for eight." He brought him [and] ordained him as a Rabbi and handed him over to Rabbi Shimon ben Issi ben Lakonya, his mother's brother. Every day he [the son] would say: "I am going to my town." He said to him: "They made you a sage, and they created a gold cloak over you, and called you 'Rabbi,' and you say, 'I am going to my town?'" He said to him: "I swear, this [thought] has been abandoned." When he grew up, he went [and] said: "This voice is similar to the voice of Rabbi Elazar the son of Rabbi Shimon." They said to him: "He is his son."

On the other hand, it is not always true that the offspring of scholars turn instinctively to scholarship. Elazar's son (I wonder why we did not hear of him earlier during his father's travails) has turned from Torah, or perhaps he never took it up; he enjoyed the life of the profligate. He has become a successful gigolo. Ironically, it is the son whom the father, Elazar, would have found in the tavern at the fourth hour and condemned as a thief. Rabbi, Elazar's colleague and rival, undertook to redeem the scholar's son. He arrived in the town where Elazar had lived (and apparently died), and wonders if this Elazar has left a son. Obviously, Rabbi has lost touch with his colleague over the years. When the son is pointed out to him, it is not the scholar, but the rake that Rabbi discovers. He is appalled, but out of respect for his friend and for scholarship, Rabbi commits himself to return Elazar's son (also unnamed here) to Torah. I wonder why? Perhaps it is the value of learning that Rabbi means to instill in Elazar's son and therefore, to perpetuate wisdom. But then, why is it Elazar's son and not any other child of a father who Rabbi would reclaim? It might also be true that honoring one's father (Honor thy father and thy mother!) is to continue not necessarily in his ways but in his manner. For Elazar's son, that would be the way of scholarship; Elazar was certainly an independent thinker. I recall that the good student presents to the teacher the most difficult objections. But perhaps these two reasons are identical: learning that does not continue to the next generation is meaningless. Rabbi means to recover the soul of Elazar's son.

Teachers do this. A good teacher changes lives. This story suggests, however, that it can't be done with grades and diplomas and degrees. After all, Rabbi ordains Elazar's son, gives him a golden cloak and awards him the title, "Rabbi." Elazar's son is given all of the accoutrements of the scholar, but he refuses to turn from his life of profligacy and commit to a life of learning. He is tutored by Rabbi Shimon ben Issi ben Lakonya, but he remains dissatisfied and daily threatens to leave his learning and return to his prostitutes. Finally, I suppose, the son knows no motive for learning outside the artifacts of it.

Elazar's son is not unlike so many of our children in the schools. There is no meaning to their learning outside of the grades they receive and the titles they earn. They would, at the school bell, run frantically to town.

What eventually leads to the reformation of Elazar's son? Rabbi Shimon admonishes him that he has gained great honor from his learning; why would he now abandon it? It is not an uninteresting question at this point of the discussion. If increase of knowledge increases sorrow, then who would willingly forego the pleasures of the flesh for the sufferings that accrue from learning? I do not think the answer is forthcoming in Talmud. In this discussion concerning what "in Torah" means, it is clear that it is wisdom that is sought and not status; to be "in Torah" is not tied to status, but what exactly does it mean to be "in Torah?" In any case, it is this reprimand that appeals so strongly to Elazar's son: he puts aside all thoughts towards the carnal pleasures of the town, and returns to study and scholarship.

When next Rabbi visits the town, Elazar's son has grown up. Though Rabbi hears the voice of the father, it is the son he sees.

✡

He [Rabbi] quoted concerning him: "The fruit of the righteous is a tree of life, and he who acquires souls is wise." "The fruit of the righteous is a tree of life"—this is Rabbi Yose the son of Rabbi Elazar the son of Rabbi Shimon. "And he who acquires souls is wise"—this is Rabbi Shimon ben Issi ben Lakonya.

When he [Rabbi Yose] died they brought him to his father's cave. There was a snake encircling the cave. He said to it: "Snake, snake, open your mouth, and let the son enter [and be next] to his father!" It did not open [its mouth] for them. The people thought to say that the one was greater than the other. A [heavenly] voice went forth and said: "Not because the one was greater than the other, but rather [because] the one endured the suffering of the cave and the other did not endure the suffering of the cave."

Rabbi Yehudah ha-Nasi quotes Proverbs 11:30 to describe Rabbi Yose, son of Rabbi Elazar, son of Rabbi Shimon. I recall that earlier this latest progeny was unnamed; clearly one gains name and presumably, agency when learning is acquired. Since Elazar (and Shimon) were such scholars, then Yose as their fruit now becomes a source of strength and wisdom. Intelligence seems to run in the family here. What with all the talk these days of the power of our genes and the inheritability of intelligence, a myth that refuses to die, one could wonder why the educational conservatives haven't looked more carefully to the Talmud to give credence to their position. It would seem that on the surface, at least, the Rabbis come down in these stories on the side of nature. After all, like father like son. The difficult but brilliant Elazar is the son of the great Rabbi Shimon; Yose is the brilliant progeny of Elazar. However, the story declares that it is not inherited intellect that is the fount of wisdom; rather, wisdom resides in the work of activating that intellect. That is, what is counted as wise is not the student, Yose, but the teacher—he who acquires souls! The fruit may become the sustaining tree, but the wise is the nurturing teacher.

I think it is very important to come from an environment that supports learning. My children, though hardly wealthy, are privileged children. Emma and Anna Rose have spent their young lives surrounded by books and music and art because we have as adults filled our house with these treasures. They are the stuff of our lives. Our rooms—their rooms—are filled with volumes of all varieties. Our children have been read to from the first moments of their lives. And read to continuously. The variety of print material by which they have been nurtured from birth is immense. There is no place to go in our house where they will not be surrounded with bound copies of the printed word. My children are excellent readers.

But what of the children who enter our classrooms and who do not have this luxury as a natural resource and whose parents do not have time or resources or inclination to create an environment of such literacy? How do we care for their souls? How do we offer to these children the resources my children acquired by the luck of their birth?

For thousands of hours my children have sat in our laps—and in the lap of a dear, dear woman, Julie Friberg, and were read hundreds and hundreds of books (often read the same book hundreds of times). I am certain that it was as much the intimacy of the experience as the quality of the book that awakened the affection for books that both Emma and Anna Rose display. Oh, they are not perfect readers, and we have struggled, and will continue to struggle with both about their reading: it is after all, in part, *our* lives in *their* reading for which we tender concern and care. But the children themselves never doubt the wonder of reading or the value of learning. They value books and what occurs in the reading process. They know that to read is to be held warmly in the arms of someone who loves you; to read is to be held secure. For the children who did not have this experience, we must create the experience: each child deserves hours and hours of what I will call "lap time." They must be allowed to sit in the center of an adult and be held and be read to for hours and hours and hours. The reading must be personal and it must be warm; the reading must be intimate and it must be important. The reading must enchant and it must satisfy. The reading must be love.

Teaching has never been a well-respected profession in the United States. Rather than be called teacher, many of us would be known as scholars. Many of us move to the university level where scholarship receives greater support and time is offered for its development. The teacher-as-scholar is a troubled concept in the United States. To spend one's life with children is ridiculed in this society: "those who can, do, those who can't, teach" is a slander infecting our social body. At the university level there is always the conflict between the demands of teaching and the demands of research. Often they are viewed as antithetical challenges; the luxury of scholarship is denigrated for the real work of the classroom. The scholar lives in the ivory tower while the teacher struggles with the students.

This Talmudic tale suggests that wisdom inheres in the teacher who gathers students to learning. I think we all have our Rabbi Shimon: I have named Mr. Matienzo and Mr. Bartul. There have been others: Doctors Matthew Wise and Edward Guereschi, to name just two. I feel myself fortunate. Their wisdom derived not from the quantity of their knowledge, but from their capacities to teach. It is not by the length of their curriculum vitae that they are to be measured; rather, they are to be known by the quality of their classroom work. They have opened their laps to the student, as it were. It is clear that the great teacher must be a great learner: but the wise is she who can attract others to scholarship. These are teachers.

When it came time to bury Rabbi Yose, he was denied entrance to the cave because he had not endured the suffering in it that his father, Elazar, and his grandfather, Rabbi Shimon ben Zakkai, had endured. It would seem again that the Rabbis are troubled and conflicted by the experience of suffering: here they suggest that if there are rewards for suffering, they do not occur in this or the afterlife.

✡

Rabbi happened to come to the place of Rabbi Tarfon, [and] said to them: "Does that righteous man who used to 'bury his children' have a son?" They said to him: "He does not have a son, [but] he has a grandson, and every prostitute who is hired for two hires him for eight." They brought him before him, [and] he [Rabbi] said to him: "If you repent, I will give you my daughter." He repented. There are [some] who say: He married her and divorced her. There are [others] who say: He did not marry her at all, so that they should say that on account of her he repented.

The Talmud here, as in several of the previous stories, portrays men as profligate and vain. Preferring the tavern to the shul and the house of prostitution to the house of study, men must be attracted to learning by promises of accompanying physical pleasures. It is an artifice we often employ with children, I think. We say, if you eat your peas, you may have dessert. We adjure: if you clean your room, you may go out and play. We promise: for every A on your report card we will give you five dollars. I am not overly critical of this practice. The realm of study and right behavior is very abstract and its promises so vague and immaterial that unless early schooled in study and etiquette—as were Elazar and Rabbi and, I hope, my children—or unless immersed in the pleasures of learning—as were Resh Lakish and Elazar's son and, I hope, my children—there appears little to capture one's interest, compared, let us say, to the physical attractions of the world. Even then, as for Rabbi Tarfon, who was a scholar who would swear by the lives of his children saying, "May I bury my children if. . . " there is no surety that the appeals of learning will continue to attract. For example, where I live in Wisconsin the fourth grade social studies curriculum concerns the state's history. Emma brings home near-perfect scores on all of her worksheets and tests. But when I ask my daughter why she is studying Wisconsin history, she responds that that is what her teacher teaches. I ask her if it is important to know about the environment in which she lives and she shrugs her shoulders. When I ask her what she thinks of Wisconsin now, having learned its history and geography etc., etc., she responds, "It's ok," and then asks if she can now go and read her book. But her book is not about Wisconsin and it is not historical. You see, she lives here and Wisconsin is there.

In education, we often complain about the curricular separation of the practical from the theoretical. We express concern over the conflict between the child's desire for satisfaction in the present and our uncertain promises of future reward; we are troubled by the conflicted appeal to the child of the physical attractions of the world and the elusiveness of the abstract in our educational process. We insist that the education in schools be practical. Personally, I find this debate specious: the very existence of the structured school and its clear-cut curricular stream attests to the purely academic nature of knowledge and its existence apart from the world of the practically living. What is learned in school rarely has any immediate relevance outside the requirements of the assessment tools of the educator. How could it be otherwise: school is separate from life. No matter how relevant we mean to make the curriculum, it is always an ideal relevance to which we refer, a reference to a world outside the lives of the school and of the students. The real question facing the schools should be how to create a curriculum that satisfies both the body and the spirit of our children not separately but as inseparable. Wisdom without deeds is weak; deed without wisdom is dangerous. Rabbi Hanina said, "He whose works exceed his wisdom, his wisdom shall endure; but he whose wisdom exceeds his works, his wisdom will not endure." I think Rabbi Hanina has spent time in our public school classrooms.

Of course, there are those who would say that doing without thinking is dangerous: but I do not think that Hanina suggests here that acting is ever done without some wisdom—that wisdom derives always from the Biblical directive *to be holy*. No sooner are the people of Israel freed from bondage in Egypt then God speaks (Exodus 19:6) to them saying, "You shall be to me a kingdom of ministers and a holy nation." The charge here is to engage in deeds founded in wisdom—the covenant. The text of Torah offers blueprints for the achievement of holiness—it is God's way. Our works aspire toward that goal, and our wisdom must lead us toward it. When wisdom exists without deeds, it is an ideal world from which people living their daily lives are necessarily absent; what good is wisdom in the absence of deeds? Every event evaluated on the basis of a holiness reforms the world; no matter how rigorous we set our curricular standards, we will fail to educate our children in the schools where there are no real deeds.

What is wisdom? The redactors of the Talmud set as its purpose the explorations of the details of daily life based in the directive that holiness in the image of God is the goal of human action. "I am holy," says the God in Leviticus, "so you shall be holy." The scholars of the Talmud examined the questions that arise during the exercise of the diurnal deeds of people living their ordinary lives; these scholars struggle to discover what behavior would be consistent with the Torah's command to be holy. The Rabbi's desire to create a society in which the daily lives of its people achieve a standard of behavior that is spiritually ennobling, underlies each word of their debate. Wisdom pursues holiness not in thought alone, but in deed. The issues in which Talmud deals always concern right behavior consistent with the words of Torah. Textual interpretation becomes a social principle. The existence of teachers is proof text that learning is requisite to a better world, even as the existence of the prophets in the Torah are an argument that the world is not yet holy. It is the charge of both teachers and prophets to argue for a better world and not merely a smarter one. Hence, the concern of Rabbi over the profligacy of Tarfon's grandson. And hence, the concern of the Biblical prophets for whom daily life exists on a level far from holy. Isaiah 59:9–11 argues that "justice has become distant from us and righteousness has not reached us; we hope for light, but, behold, there is darkness; for brightness, but we walk in deep darkness . . . We hope for justice, but there is none; for salvation, but it is distant from us." In prophetic desire, we hope that our children will learn in school how to repair the world so that justice and righteousness would be brought near and that we might walk in light and not live in darkness. On the news I hear horrifying reports of the war in Iraq, of bitter fighting between Palestinians and Israelis, of the torture of a Haitian immigrant by an officer of the law, of the massacre of fifteen people in Littleton, Colorado. What does algebra have to do with any of this? Or Wisconsin history? If our society continues to assess its educational system by the attainment of high measures on standardized tests of externally constructed curriculum mandates, we will not consider a curriculum that ensures a more compassionate and just order. If our curriculums are written to achieve rigorous academic standards, then these may not also be those standards that will ensure justice and freedom. If we decry the decline in test scores of our children as harbingers of a declining social fabric, we are blind and deaf to the unraveling of our children's lives in the social milieu into which we have brought them.

We are in schools much in need of the prophetic imagination: perhaps it is the prophetic tradition to which teachers might adhere rather than that to that of the statisticians. On this theme Abraham Joshua Heschel

(1959:125) is eloquent: "Like a scream in the night is the prophet's word. The world is at ease and asleep, while the prophet is hit by a blast from heaven. No one seems to hear the distress of the world; no one seems to care when the poor is suppressed. But God is distressed, and the prophet has pity for God who cares for the distressed." The prophetic imagination that should drive curriculum in the schools should be, says David Purpel (1988, 85), "concerned with an alternative society, one with sacred dimensions," and not with the competitive dynamics of the world economic markets. Our education should be directed toward raising the level of the sacred and not of test scores. For the Rabbis of the Talmud, it is learning that will lead to raising standards. However, it is not the products of learning—the test scores—that they value so much as the exercise of that learning, the conversation and debate. For the Rabbis, the purpose of learning lies in the continual quest for holiness. All of the value judgments by the Rabbis are based in this standard. The spirit and the body are irrevocably linked in this learning: finally, without action, learning is meaningless, and action derives from the ethics of a daily life founded in holiness. What we fail to offer in schools is the excitement that derives from meaningful learning itself! What we fail to offer in schools is the absolute connection between the daily and the sacred! What we fail to offer in schools is the prophetic voice that demands justice, compassion, hope, energy, and engagement! What we fail to make use of in the schools is the wonderful questions children make about the world and, of course, the wonderful questions children inspire in us by their questionings and behaviors! What we fail to be concerned with in the schools is the matter of our children's spirits! Rather, in our schools knowledge is treated as if it were ideal, free of value, and freely available to all equally.

On April 20, 1999, two students sat in their classroom and smiled conspiratorially at each other. They might have been in mathematics or social studies or English or Art. They might have sat in rows, or in circles, or on the ground. They might have had their texts and their notebooks in front of them; in their folders they might have placed tests they have passed. They were due to graduate. And for an entire year these two have been plotting the massacre of scores of their classmates and teachers; for an entire year they have constructed materials that will physically destroy the school edifice itself. For a year they have been stockpiling weapons they will use to murder their community. Tomorrow they will execute their plan and then they will execute themselves. But no one knows what they plan. No one knows! They have been able to keep this monstrous plot a secret for a whole year. *And yet the President of the United States cannot conceal a simple blow job.* There is something desperately, even tragically amiss.

I think it is this concern that motivates Rabbi to seek out the grand-son of the man who used to swear by his children. I think it is this con-cern that leads Rabbi to return the student to learning. I think Rabbi means to return Tarfon's grandson to other deeds than prostitution.

The conflict from the beginning has been between the separate appeals of the material and the physical world, the world of study and the world of daily life. But unlike the Platonists, who discount the physical world as shadow and unlike the materialists, who discount the idealist world as ephemeral, the Rabbis of the Talmud recognize that an intimate relationship between the physical and the spiritual must be fashioned. Rabbi Elazar's son had been turned from his wantonness by study; earlier Resh Lakish had been enticed to study by Rabbi Yohanan and the prom-ise of his beautiful sister and himself became a great scholar. Now, the grandson of the man who used the oath, "May I bury my children" is seduced away from his sexual debauchery to the adventures of study. I am sorry that the Rabbis invest the limits of the physical in women, and so it is they who must be abandoned, or must be made to suffer, so that men may study. I think the stories suggest that the turn to study creates con-flicts in the physical world that must be resolved if the effort of study—the achievement of holiness—will be fulfilled.

For me, it is not my sense of these stories that the student may callously expect others to suffer on his behalf. Rather, these stories reflect the serious conflict the Rabbis experience between the claims of study and the claims of daily life. The Rabbis cannot ignore the intimate relationship that exists between the spiritual and the physical. And I am also relieved that there is in Talmud support for my efforts in classroom to seduce students to study; I am also pleased to note that my struggles are not new ones. It is not always easy to convince others of the value of study. And sometimes exter-nal rewards do attain a measure of success; we hope that students will refuse to accept them to avoid rumor that they work only for material honor. We hope that they will give those material pleasures up when learn-ing itself becomes its own reward. Study has a different trajectory than the life of material gain. The Talmud suggests that the scholar's life must draw a balance between the physical and the spiritual, between the profane and the sacred. The existence of the world depends on this balance.

✡

And why did he do all this?
    As Rav Yehudah said in the name of Rav, and some say [that] Rabbi Hiyya bar Abba said in the name of Rabbi Yohanan, and some say [that] Rabbi Shmuel bar Nahmani said in the name of Rabbi Yonatan: "Whoever

teaches his fellow's son Torah will merit to sit in the Heavenly Academy, as it is written: 'If [Israel] repents, I will bring you back [and] you will stand before me.' And whoever teaches the son of an ignorant person Torah, even if the Holy One blessed be He, has issued a decree, He will cancel it on his account, as it is written: 'And if you bring out the precious from the vile, you will be as My mouth.'"

The teacher's motivation is the subject of this lovely story, but I think the Rabbis of the Talmud understand the role of the teacher far differently than does the American educational establishment. Here, in America, one's status depends in large measure on the achievement of students on standardized tests. There, in Talmud, when someone asks why Rabbi had gone to all that trouble to entice Tarfon's wayward grandson to study, someone answers that whoever teaches his fellow's son will earn a seat in the Heavenly Academy. Of course, since everyone is someone's son or daughter, then the promise made here is that the teacher's work will earn him/her a place in the Heavenly Academy—the Teacher Hall of Fame, so to speak. News of this assurance might make irrelevant the absurd search for Teacher of the Year. Here, the Talmud argues that all teachers—"whoever teaches his fellow's son"(sic)—regardless of immediate success are to be awarded a place in the Heavenly Academy. In this regard, it is the engagement with the subject that defines the teacher rather than, in our present educational system, the teacher merely declaiming the subject. Nor need Torah be here understood in its narrow religious sense; rather, Torah is an attachment to history, even to local custom, as it were. Torah, I recall, is the instruction manual about how to be a holy people. To teach Torah is to ensure that the holiness that is God of the Torah lives amongst us, even if in the practice of our daily lives we consider that God's face remains hidden. Only the study of the word of the text gives us confidence that a God exists; that text, the Rabbis tell us, was the blueprint God read when God created the universe. Rabbi Elazar said that, "Where there is no Torah there is no culture; where there is no culture, there is no Torah." The purpose of pedagogy is the engagement with Torah. The purpose of pedagogy is the continuance of culture. The purpose of pedagogy is the repair of the world.

And so, Rabbi's pedagogical intent is justified by quoting Jeremiah 15:19. The relationship between the prophet and the teacher is made explicit in this reference. To my mind, it is an interesting passage to which the Rabbis here refer. In chapter fifteen, Jeremiah complains to God that his prophecy has made him the subject of persecution and exile; Jeremiah bemoans his seemingly pointless sufferings and his meaningless prophecies. Jeremiah regrets having ever been born! He complains: "Why has my pain become everlasting and my wound acute. It refuses to be healed. You have become like a disillusionment to me, like unfaithful waters." We are in familiar terrain: the world of the suffering. Jeremiah suffers and his anguish confuses and displeases him. He complains to God. Unlike Elazar, Jeremiah neither invites his sufferings nor wills them away to return to study; unlike Rabbi, Jeremiah's sufferings neither derive from an event, nor are they borne without recrimination; unlike Rabbi, Jeremiah does not appreciate his sufferings as precious. But then, Jeremiah as prophet-teacher must speak before a class of students whose immersion in immediate sensual pleasures make them deaf to his warnings. Jeremiah accuses God of tormenting him unjustly by demanding he be a prophet to those who would not listen. Jeremiah complains that he speaks to the vain in vain.

God, however, will have none of this caviling; God demands an apology from Jeremiah for his complaints: "If you repent I will bring you back . . ." God tells Jeremiah that despite his depression and hopelessness, he must every day behave as if the world would repair. More, God promises redemption to Jeremiah if he can also redeem the people; if he teaches the people he will himself be saved. God offers the teacher salvation for achieving the salvation of others. The one who saves the Other saves the world. It is the teacher who has the power to turn people from their ways and return them to the goal of achieving holiness. Learning is, as I have long suspected, redemptive. But first, the story suggests, the teacher will feel mocked and neglected. Recall that Lakish tried to retrieve his clothes on the banks of the river and run, and Elazar's son and grandson head first for the brothels. In the end, however, "they will battle you, but they will not be able to overcome you . . ." In referring to the prophet, Jeremiah, as proof text, Rabbi looks to Jeremiah as the model of the teacher.

✡

Since the beginning of the common school movement in the first third of the nineteenth century, schools have been touted as redemptive. Education was proclaimed as a way to produce a society capable of democratic rule and one worthy of that democracy. The struggle over the American curriculum has forever been between ideologies that espouse a specific social vision as the only correct one. I do not intend to add my voice to that din. Today's educational debate over the role of standards and national curriculums merely extends that debate, but to my mind does not offer much hope. I am, however, intrigued here by the Talmudic discussion: it argues that the teacher is s/he who returns the people to Torah—to study. Rather than look to the student's development of any specific product, the Rabbis measure the student's engagement to process as the criteria for the performance of the teacher. Education in Talmud is redemptive; it is a turning. Perhaps those involved in education already know that—though we will see that even scholars lose sight of the purpose of study. But Talmud argues here that the teacher who teaches the son of an ignorant person can save society from its self-destruction. (Doesn't this sound remarkably like the elitist curriculum of the nineteenth century that meant to inculcate the children of immigrants into the white male Anglo-Saxon culture?) God promised Jeremiah that if he would inspire the return of the people to the ways of holiness, he would bring redemption upon himself; God assured Jeremiah that once the people had begun to understand the prophet's argument for study, they would continue to look towards Jeremiah for guidance and wisdom. God tells Jeremiah that though times will continue to be hard, Jeremiah will finally succeed. God tells Jeremiah that he need not immediately succeed; indeed, God warns Jeremiah that he will not realize any success in the near future. But God does tell Jeremiah that he will gain his strength and power not from the *achievement* of success but from the *engagement* in effort.

And that is why Rabbi has done all of this for Rabbi Tarfon's grandson. The Greeks, I recall, gave us Sisyphus; the Jews gave us Rabbi.

✡

> Rabbi Parnakh said in the name of Rabbi Yohanan: Whoever is a scholar, and his son is a scholar, and his son's son is a scholar, Torah will never again cease among his descendants, as it is said, "And as for Me, this is My covenant . . . they shall not depart from your mouth nor from the mouth of your descendants nor from the mouth of your descendants' descendants, says the Lord, from now and forever." What is [meant by] "says the Lord?" The Holy One, blessed be He, says, "I am a guarantor for you in this matter." What is [meant] by "from now and forever"? Rabbi Yirmeyah said: "From now on the Torah returns to its own home."

Rabbi Parnakh remarks, in the name of Rabbi Yohanan, that if the Torah is studied for three generations it shall never be lost from that lineage. I am glad that the Rabbis fall on the side of nurture in the debate over which has greater influence in the educational potential of the child—the total environment or the biological gene-pool. In our modern era when even the propensity to smile is attributed to a specific gene, I am comforted by the Rabbis' beliefs regarding the efficacy of human purpose and influence. They argue that, situated in an environment in which study is practiced regularly and with fervor and purpose, embraced by the practice of study and seduced by its beauty, the desire for study will become tradition and will forever remain within the family group. It is important study, I think. As I have said, the Rabbis of the Talmud taught that when God created the world, God looked within the Torah for the blueprint. Within the Torah one finds the world. The study of Torah is the study of human existence.

This is in sharp contrast to the Olde Deluder Satan Law passed in Massachusetts Bay Colony in 1647. There the magistrates meant to protect the truth of Scriptures by insisting that children be taught to read it properly. For these governors, meaning existed prior to the words and could be discerned properly by reading rightly, looking as it were, through a glass darkly. The law intended that every student be taught to read so that "the true sense and meaning of the original" be known. I believe that this represents a form of pedagogy often practiced in our classrooms. There we are taught to look for correct answers. But this pedagogy is not endemic to the scholarship to which the Rabbis refer as may be clearly evident from their own extended yet often unresolved debates concerning the meanings of texts recorded in Talmud. In a holy text— the Torah—every word is crucial and thus, nothing for the Rabbis is unproblematic. For example, Rabbi Parnakh states unequivocally that the tradition of scholarship is steadfast in a household in which study has been passed through three generations, and he supports his opinion by quoting the canonical book of Isaiah 59:21. One of the Rabbis asks very plainly "What is meant by 'says the Lord'?" What a remarkable question! What could be simpler than these three words? Children are early taught that this formula indicates a quote, names the speaker and in what register the statement is performed. And yet the Rabbi's query calls this simple statement into question and creates an allusive web that opens the text to multiple meanings. For this simple common phrase is packed not merely with meaning but with force. Words are not mere things but are powerful acts.

I appreciate that the Rabbis keep returning to the Prophets for their proof texts in their discussions of teachers and scholars; for an answer to this seemingly obvious question the Rabbis again turn to Isaiah. The link is again made between the scholar and the prophet. Prophets and teachers, the Rabbis suggest, have much in common. "The prophetic voice speaks most directly," David Purpel (1988:81) writes, "to issues of justice and righteousness; it is a voice that not only roars in protest at oppression,

inequity, poverty, and hunger but cries out in pain and compassion. It is
a voice that haunts us because it echoes our own inner voice partly,
which speaks to our impulse to nourish and care for others." The
prophet's voice might be understood in the voice of the teacher. In the
verses referred to in Isaiah (59:16), the prophet speaks of God's wrath
and disappointment that "there was [in the world] no worthy man" to
set for others an example. As a consequence, God must then rely on
God's own benevolence to redeem the Israelites. Despite God's anger and
vengeance, Isaiah promises the people that a redeemer will come to Zion.
Isaiah's text reads: "And as for Me, this is my covenant with them," said
Adonai, "My spirit which is upon you and My words that I have placed
in your mouth will not be withdrawn from your mouth nor from the
mouth of your offspring nor from the mouth of your offspring's offspring," said the Lord, "from this moment and forever." Thus it is that
the Rabbis suggest that learning that remains throughout three generations will never cease from among the descendants.

And it interests me that, whoever offers Isaiah as proof text, carefully
omits the words "with them," which would specify a clearly historically
situated population to which God's words would apply. The omission
ensures that God's promise remains throughout time. It is interesting as
well that another Rabbi offers *this* interpretation of the particular words
in question: "'says the Lord'," the Rabbi says, "means that God is the
guarantor in this matter." That is, the phrase "says the Lord" is not merely
a textual device but a contractual formula. It is an indication that Isaiah
is not making this stuff up—the guarantor of redemption is God! In this
particular case in Talmud, it is God who guarantees the permanence of
scholarship in a family whose tradition is scholarship. The Rabbis' interpretation suggests to me that scholarship is divinely protected.
Scholarship is sacred. I like that. I like as well the Rabbis ascription to the
simple phrase, "says the Lord" the power of an oath. That is, when God
speaks, it will be as it is spoken. If God is holy and therefore we should
be holy, then when we speak, all should be as we have spoken. What an
ethical principle is there!

Then a second seemingly irrelevant question is asked: "what does 'from
now and forever' mean?" Isn't that, too, obvious? Clearly another agenda
is at play in these simple questions. The Rabbis are talking about the persistence of scholarship that extends over three generations—a relationship
established between the text and the reader as a result of the tradition of
study. And so I believe that the Rabbis inclination is to read this phrase
with broad sympathy: within the comfort of the scholar's home the text
will always feel sustained. The text has become a living document that

requires the scholar to ensure its well-being and good feeling. Gatsby, I recall, never even cut the pages of the books that lined his shelves.

The discussion that had begun by questioning what "in Torah" meant, has now turned to questioning the meaning of education. "In Torah" has not yet been defined—I am not yet certain it ever will be: the Rabbis continued discussion reminds me that the text is always in need of interpretation. Its meaning is "not in heaven;" all depends on local custom. It is study that is important and it is the work of the teacher that promotes that redemptive activity. It is the work of the prophet: immediately following the verses of Isaiah comes the following joyous shout (60:1): "Arise! Shine! For your light has arrived and the glory of God shines upon you. For behold, darkness may cover the earth and a thick cloud [may cover] the kingdoms, but upon you God will shine, and God's glory will be seen upon you. Nations will walk by your light and kings by the brilliance of your shine." I would like such a classroom.

God promises the people redemption; here, the Rabbis argue that redemption will come in the form of education and be facilitated by the teacher.

✡

> Rav Yosef fasted forty fasts, and [in his dream] they read to him: "They will not depart from your mouth." He fasted another forty fasts, and [in his dream] they read to him: "They will not depart from your mouth and the mouth of your descendants." He fasted another forty fasts. He came and [in his dream] they read to him: "They will not depart from your mouth nor from the mouth of your descendants nor from the mouth of your descendants' descendants." He said: "From now on I do not need [to fast]; the Torah returns to its home."

I am envious of the clarity and directness of Yosef's dreams. Their meaning is transparent and their results direct. What he studies in his dreams mirrors his actions and heralds his future. Though I, too, hold fast to my dreams, I am less confident of their coherence and simplicity. I know my dreams are meaningful, but I am not always certain when or how they may achieve fulfillment. The Joseph of Genesis dreamt and thought his dreams were real; he had to learn that though dreams are purposeful, we do not always know how our lives will realize the purposes of our dreams.

In this story, Yosef's dreams are connected to his fasts. Perhaps while he is studying, he refrains from eating as a way to intensify his ability to fix his attention. His fasting, and therefore his studying, inspires his visions. His studying and his fasting and his dreaming fulfilled him. Once, twenty years ago I attended a concert by the Grateful Dead. I did not eat but had eaten. The band was playing "Tennessee Jed." I have always known this song to be about the dream to return to simpler ways and simpler days. I remember standing in the huge concert arena. Suddenly, awash in the music and afloat in an LSD happening, I had a vision of myself walking effortlessly toward a small wooden farmhouse surrounded by a white picket fence in front of which daisies grew and on which bursting scarlet rosebushes clung. In the vision, I felt an enormous peace as if my body had no weight and knew no threat. I knew I was whole; I knew I was home. It was a wonderful dream I have held vividly in my memory these twenty years. After the concert, I returned to my New York City studio apartment and left my vision behind in the concert hall.

I live now, twenty years later, in rural Wisconsin in the midst of part of that vision. I live on a mere four acres; but to the east of me are fifty acres of someone else's farmland and to the north lies an enormous open field, also owned by someone else. To my south is someone else's hobby farm on which cows and horses regularly graze. Our nearest neighbors are several hundred yards through a narrow wood to the west, but in the spring and summer the thick green foliage obscures our view of their house. I tell my friends that to a Jew from New York's Upper West Side, I think that I live on the Ponderosa. I ride my lawn tractor as did Adam Cartwright his horse. I ask my children to call me "Paw!" Oh, it is no white picket fence and we plant no daisies or roses, but I have achieved some peace and the home I dreamed unawares twenty years ago. My children live here. I study here, and especially so during this sabbatical year.

Each series of forty fasts ensures Yosef that his scholarly tradition will be passed on from his generation to the next. After each forty-day fast, his dream assures him that his words will not depart from either his mouth or the mouth of his descendants or from the mouths of his descendant's descendants. After one hundred and twenty fasts, Yosef halts his fasting because the reading of Isaiah in his dreams assures him that the scholarly tradition will endure forever within his family.

I would have the same assuredness. The academic tradition has been my life, and books and study have been as integral to me as the air I breathe and the food I eat. My learning has never *not* enriched my life, and it still promises always the hope for tomorrow. Learning is my daily prayer and study my sacred space. I know the words will not depart from my mouth. But I want to know that the words will not depart from theirs as well. Perhaps it is the legacy I can give my children. And so I struggled with myself—and sometimes with them—to make reading critical to their lives; I have struggled with myself, and sometimes with them, to make reading wonderful to their lives; I have struggled with myself, and sometimes with my demons, to let them be children, whatever I imagine that might be. And I want to be assured that the words would not depart from their lives.

I have always worked down here in the basement. I write and read in the early morning hours before the household awakens and then go upstairs to prepare breakfasts, lunches, and assist the sun's rise. For the past several weeks when I have gone upstairs the light is already turned on in Emma's room. I poke my head in quietly. It is 6:45 a.m. She is still in bed. She turns to me and smiles: She has been awake since 6:00 a.m., she tells me. She has been reading, she informs me. She turns from me and back to her book. I think the words will not depart from her life.

✡

When Rabbi Zera went up to Eretz Israel, he fasted a hundred fasts so as to forget the Babylonian Gemara, so that it should not confuse him. He fasted another hundred [fasts] so that Rabbi Elazar might not die in his lifetime and communal cares fall upon him. And he fasted another hundred [fasts] so that the fire of Gehinnom should not have power over him. Every thirty days he would examine himself. He would heat the oven, go up and sit inside it, and the fire would have no power over him. One day the Rabbis cast an eye upon him, and his legs were scorched, and they called him "the short one with the scorched legs."

Yosef fasts to remember; Zera fasts to forget. It is the same hunger, nonetheless. Perhaps remembering and forgetting are inspired by similar pains. And perhaps these stories suggest a surer connection between remembering and forgetting: is remembering the attempt not to forget, or is forgetting the struggle not to remember? Torah presents an interesting connection between the two processes. Deuteronomy 25:17–19 admonishes: "Remember what Amalek did to you, on the way when you were leaving Egypt, that he happened upon you on the way, and he struck those of you who were hindmost, all the weaklings at your rear, when you were faint and exhausted, and he did not fear God, It shall be that when Adonai, your God, gives you rest from all your enemies all around, in the Land that Adonai, your God, gives you as an inheritance to possess it, you shall wipe out the memory of Amalek from under the heaven—you shall not forget." That is, the Torah demands a specific memory, what Amalek did to you, while it demands a certain forgetting, the memory of Amalek, and then reiterates its demand that "you shall not forget." I wonder what is to be remembered and what forgotten? How achieve the first without failing the second? That is, in this verse from Deuteronomy, God commands that I forget and remember simultaneously. Freud argued that nothing is really forgotten, but that everything is retained in a manner that merely *simulates* forgetting, hence, the experience of repetition. In the process of transformation, a symbol takes the place of that which would be forgotten. Art, then, is understood as a product of forgetting. Art is remembering in displacement. Might the Torah be understood as anticipating Freud: memory as the production of the symbol? A symbolic memory? Indeed, in Torah, the Amalekites come to personify evil in a wide variety of instances.

But Freud further argued that nothing is, in fact, remembered, and that all memories serve as screens for the memory of desire, that which can (or must) never be achieved and therefore, must be forever forbidden. The memory is a forgetting, and art is the medium of the amnesia. Might the Torah be offering some relief from my own repressed desires by portraying an identifiable (remembered) source of malevolence in the world—Amalek? The evil is out there and not in here!

On the one hand, the verses command that Amalek's betrayal be remembered, but on the other hand, the verses adjure that we wipe out the memory of Amalek from under the heaven. And then, to reinforce the seeming contradiction, the final verse adjures us "do not forget." How might I consider wiping out the memory of Amalek from under the heaven and yet not forget Amalek's treachery? Can I remember the deed and not the doer? Is evil not to be always embodied? If I remember only the deed, would it not be only the victim that I recall? In this type of remembering, evil becomes merely a malevolent force in the universe and as such, attenuates the power of resistance and the sense of affront that the evil calls forth in me? If evil is to be recalled in such a disembodied manner, might it not suggest that the immanence of evil is permanent to the world? Perhaps that is the point: evil exists not in the aberrant but in the everyday, resides not in *a* body but in *all* bodies. Evil is a choice and not a fate. Wipe out the memory of Amalek but remember what Amalek did; wipe out the face of evil and remember the evil itself!

But though I am commanded to remember the treachery of Amalek but not Amalek himself, the perpetual reading of the Torah is a constant prod to the remembrance of Amalek. As long as the Torah is studied— and that is always—then Amalek cannot ever really be forgotten, even though the Torah commands that his memory be erased. I wonder why the Torah sets me such an impossible task? Why does the Torah speak in such obvious contradiction? How could the memory of Amalek ever be wiped from under the heaven as long as the story of Amalek exists in the Torah, which is forever to be read and studied?

Perhaps Amalek is the memory of evil (the deed), and perhaps Amalek is the memory of my own forbidden desires (the screen). That is, Amalek stands not only for my repressed impulses, but also for the conflict that exists between the fulfillment of purely personal joys in the engagement of study and the more complex obligations to the social. Perhaps that is what Zera's second hundred fasts reflect: as long as Elazar lives, Zera is free from communal obligations. As long as there is someone else to assume the social responsibilities, then Zera can remain in study. Zera prays for Elazar's continued health so that Zera can remain apart from the social world and continue in study. Is Zera's study associated with Amalek's treachery? Viewed in this manner, the story of Amalek might argue for the primacy of ethics in a world where the reality of evil exists and where my seclusion in study (in the basement) aligns me with Amalek. Zera has the choice to remain in study or join Elazar in social activity. In any ethical stance, evil can be not eliminated but confronted and sometimes, controlled; study prepares us for action. But without action, study is unethical. If I am not for myself who will be, but if I am only for myself, what am I? It is our choice to move in the direction of one or the other. We are our choices.

Perhaps the Torah suggests that without an ethical stance before the Other, we all have the potential to be Amalek. Our daily lives present us with conflict between either fulfilling our own desires, or satisfying the demands of others upon us. Thus, the Torah commands that we must never forget Amalek's treachery because our lives constantly present us with the opportunities to act with perfidy. The Torah commands, for example, that employees may not withhold payment from their workers at the close of work: it is treachery that betrays the worker. The Torah commands that it is treachery that cheats the hired employee, that dishonors the maiden, that humiliates the parents, that returns the slave to its master, and that steals the neighbor's livestock. It is forbidden to punish the son for the crimes of the father. The Torah suggests that this ethics is to look the other full in the face. Amalek, we are told, attacked from the rear. The treachery of Amalek, as that of the wicked child, the bad neighbor, and the greedy employer, are rendered impossible by this face-to-face stance.

Amalek attacked the Israelites when they were most weak, and he attacked them at their weakest point. Yet through the help of Adonai and the efforts of Moshe, Aharon, and Hur, the battle was won. The command that closes these verses is "do not forget." I think we must always recall that enemies have greatest access to us at our weakest moment and at our weakest point. Amalek, we are told, because he did not fear God, struck in this cowardly manner. Perhaps it is this that the Torah reminds us not to forget: our enemies, because they are our enemies, will always attack our weakest points and in our most vulnerable moments. This memory will make us strong; it is not a memory of their evil but of our strength. It is a memory of the necessity of vigilance, of the centrality of ethics essential to our survival. Our rigor—our strength—lies in the ethical stance we become when we follow the commands of Adonai expressed in the verses preceding the commands of remembering and forgetting. That is, when I do not pervert the judgment of a proselyte or orphan I am strong; my most vulnerable point—that part which is always out of sight—is best protected by this loving concern for the Other before whom I will stand face to face. In that vigilant stance, I become an ethical person. Memory here is of my strength and not my weakness.

Of course, when the Israelites are no longer en route to Israel, then the memory of Amalek's treachery loses a great deal of its force; their achievement of the Land, it is hoped, will preclude treachery from the rear. Much in the Torah suggests that to gain the land is to achieve a security. And hence the presentment of another contradiction. The land of Israel, "the Land that Adonai, gives you as an inheritance to possess," means that there will be no rear from which to be attacked; the land of Israel signifies that the weakest points are now strong. In the inherited land, the people will have rest from enemies all round. And yet the Torah insists that the treachery of Amalek be remembered. "You shall not forget." Perhaps there is a humility expressed here: once strong, the people must not then become like Amalek. Without enemies all round, perhaps we stand at someone else's rearmost, at some else's weakest point. The Torah commands that we must always stand face to face to the Other—in that stance we protect our weakest point and comfort others in their fears.

What is it that must be forgotten? It is, I think, Amalek himself! The memory of his reality would frighten us, as do the horrors of little children, which keep them awake at night. Sometimes, it is the indistinct fears that are easier to relieve than those that derive from the clear memory of Amalek. Charles Manson is much more horrible than the monster I can imagine who accomplishes Manson's deeds, because the latter I know, is my invention. And what must *not* be forgotten? It is the memory of Amalek's *deed*: it is the evil and not the specific evildoer that I must remember. Perhaps it is that there is no end to Amalek; he need not be remembered himself for he is always here: only his visage changes.

✡

But Zera fasts as well, we are told, so as to forget the pedagogy of the Babylonian Talmud. This statement requires some explanation. After the exile ended in 539 B.C.E., many Jews returned to Palestine, but many Jews continued to live and to study in the communities they had established. There existed two chief centers of Jewish intellectual activity: Babylonia and Jerusalem. Their texts were similar, but because of the contexts, their methods were different. For example, in Palestine, where the Jerusalem Talmud was redacted, the Rabbis held that since they lived in the Promised Land, they were especially able to interpret the law accurately without the often-difficult analysis that engaged the scholars in exile in Babylonia living far from the traditional center of Jewish learning. Furthermore, certain agricultural laws applied only to the land of Palestine and were not a subject for the discussion of the scholars in exile. Finally, since in the exile the Rabbis were not authorized to rule on certain issues—these could only be legitimately decided by the Rabbis in Palestine—those scholars in Babylonia tended to engage in more open-ended and speculative conversation. The two Talmuds, the Babylonian and Jerusalem editions, represent two ways of thinking and perhaps two forms of pedagogy. Once he has immigrated to Palestine, Zera wished to forget the ways of scholarship he had learned in Babylonia.

For all of my life I have practiced Western forms of thought. I have learned and learned again that American education has been centrally founded upon Greek metaphysics. American education has always sought what knowledge is of most worth. It would seem that a great part of American education does not know philosophers before Plato. Despite the immersion of Jews in study, there are few references to Rabbinic thought or Jewish educational practices in the discourse of American education. I am and have been a teacher for twenty-five years in the public secular arena; I have not spoken as a Jew because I did not have and was not offered a language with which to do so. That is, I taught as the Judeo-Christian world (a term which is alien to me) taught me to teach; I used the texts I was given, and they were always Christian texts; I did not know there were alternatives. As Leslie Fiedler reminds me, those texts I taught as "our heritage" were really "their heritage—which is to say, Christian and therefore, necessarily anti-Semitic." I learned to speak and even to love a language that not only excluded me but gave voice to hatred of me. Now I help others learn how to become educators, and I discover I am using the language with which I was taught.

During this sabbatical, I fast a hundred fasts to forget the Western canon. Now I would like to speak as a Jew. A student once asked me:

"What do you intend to do on your sabbatical?" and I responded that I wanted to learn to think in a new way. She said, "How will you know if you've succeeded?" I responded, "I probably won't, but perhaps you will, and you will help me."

✡

Finally, Zera fasts to protect himself from the fires of Gehinnom. Ironically, it is nothing that he does that makes him finally vulnerable to the fires; rather, it is the glance of the Rabbis that renders him susceptible. But I wonder if perhaps this story is meant to suggest that the Rabbis might have looked jealously at Zera because he was able to exist aloof from the daily problems of existence to remain engaged in study. It was their look that made him vulnerable to the fire; the scar he acquired served as a life-long remembrance of this event.

Sometimes, too, it is the commonplace that I adore in these stories. Students often re-name their teachers to define their relationship to them in the classroom. It is perhaps the only way students might assert some control over their circumstances. In surreptitiously passed notes and whispered giggles, students talk about "the cat woman" and "Peeps" and "Dragon Lady" and "Lurch" I am myself sometimes referred to as "the dread Dr. Block." Or "Vinegar, Son of Wine."

✡

> Rav Yehudah said in the name of Rav: What is [the meaning of] what is written: "Who is the wise man that may understand this, and who is he to whom the mouth of the Lord has spoken, that he may declare it, for what the land perishes?" This matter the Sages said, but did not explain it; the Prophets said [it] but did not explain it, until the Holy One, blessed be He, Himself explained it, as it is said, "And the Lord said, 'Because they have forsaken my Torah which I set before them.'" Rav Yehudah said in the name of Rav: Because they did not recite the blessing for the Torah first."

This story stands at almost the center of this *sugya* and at the thematic center of my text. It portrays for me the glory and the struggle I experience in classrooms. The Rabbis have begun to explore the question of what it might mean to be great "in Torah;" the rabbis have begun to wonder what it means to be a scholar and how scholarship might be measured. What does it mean to engage in scholarship? To be a scholar? The Torah, in its Oral and Written forms, had been given to the Hebrew people at

Sinai. Since the revelation, the interpretation of its teaching has become the blueprint of a social order. But the people of Israel have not always maintained the Temple, or the scrolls it contained, nor even the teaching contained in the Torah scrolls. It is a sad history of social unrest and violence, of banishment and persecution, and of intellectual and ethical disregard. Thus it was that the knowledge of Torah precipitously declined even as the level of the social order deteriorated. But in about 539 B.C.E. by the Proclamation of Cyrus, the exiles were permitted to return from Babylonia, and the people returned to their homeland. Of course, the return was not simple. Factional infighting and tensions between the returnees and the Israelite inhabitants who had not been exiled, economic hardship, and political conflict made the task of unifying the people difficult. Attention to the texts of Judaism continued to be neglected. Certainly, the regular and pervasive practice of studying Torah was, for the most part, ignored.

Finally, in 519 B.C.E., the rebuilding of the Temple, which had been destroyed in 586 B.C.E. by the marauding Babylonians, was completed. But economic and social conditions did not improve. The people lived in absolute poverty and misery, and the people murmured and moaned. Ezra arrived in the middle of the fifth century B.C.E. as the head of a group of exiles specifically charged by the Persian king, Artaxerxes, to return the "law of the God of heaven" to Jerusalem. Under the leadership of Nehemiah, appointed governor of the province and given authority to restore the ruins, the walls of Jerusalem were rebuilt and the exiles gathered as one people before the Water Gate.

There, on the first day of the seventh month (Rosh Hashanah) in 444 B.C.E., Ezra, the priest and scribe, restored the public reading of the Torah

Then all the people gathered together as one man [sic] at the plaza before the Water Gate and they asked Ezra the scholar to bring the scroll of the Torah of Moses, which God had commanded to Israel. So Ezra the Kohen brought the Torah before the congregation—men and women, and all those who could listen with understanding—on the first day of the seventh month. He read from it before the plaza that is before the plaza that is before the Water Gate, from the [first] light until midday, in front of the [common] men and women and those who understood; and the ears of all the people were attentive to the Torah scroll. Ezra the scholar stood on a wooden tower that they had made for the purpose; next to him stood Mattihiah, as well as Shema, Anaiah, Uriah, Hilkiah, and Maaseiah on his right; and on his left, Pedaiah, Meshael, Malchijah, Hashum, Hashbaddanah, Zechariah, and Meshullam. Ezra opened the scroll before the eyes of all the people; and when he opened it, all the people stood silent. Ezra blessed God, the great

> God, and all the people answer, "Amen, Amen," with their hands upraised; then they bowed and prostrated themselves before God, faces to the ground . . . They read in the scroll, in God's Torah, clearly, with the application wisdom, and they helped [the people] understand the reading (Nehemiah 8:1–8).

It is, to my mind, a magnificent image: the whole population gathered to hear the scrolls of the Torah read in their entirety for the first time, and at one time, in very many years. The reading went on for seven consecutive days. After years of exile in Babylonia, the Israelite people had begun to re-settle in Palestine; they rebuilt the walls and rededicated Temple. They desired as well, we are told, to return to the ways of Torah. On that first day, Ezra and his Levite translators mounted the wooden platform before the entire population and begin this first public reading of the scroll. And the text tells us, that the first words that Ezra utters are not from the text of Torah but were rather, a blessing praising God's greatness and goodness. Ezra praises God for the gift of the text of Torah. As a result of the blessing, this public reading becomes a holy enactment. Holy, as Abraham Heschel reminds me, is not the equivalent of good. Genesis states that the product of the six days of creation was *good*, but that the Sabbath day God made *holy*. The products of the six days of creation are to be admired and assessed, but the Sabbath day is to be lived in awe and wonder. And so, the first words Ezra speaks are a blessing, because this public event is not merely study, but the engagement with that which is holy, with that which inspires awe and wonder. A blessing is a moment of insight, a chance of direction. Prayer raises the mundane to the level of the sanctified. Prayer is the awareness that we live amidst daily miracles and that there is more to the world than we will ever know. Heschel (1959:52) writes, "The beginning of awe is wonder, and the beginning of wisdom is awe." Prayer is an expression of awe; prayer sacralizes the mundane. Torah reading is not only study, but an engagement with wonder and awe. This study is preceded by a blessing.

Today, public reading of Torah takes place on Monday and Thursday mornings—traditional market days—on each festival day, and, of course, always on Shabbat. At each of these readings and before a word of the text is read aloud, a blessing is recited. "Blessed are you, Adonai, our God, Ruler of the Universe, Who has sanctified us with Your Commandments and has commanded us to engross ourselves in the words of Torah." This blessing raises study to the level of the sacred. To engage in study is to be brought near to the holy.

Thus, the story in Talmud intrigues me. I am a teacher; I engage in study, I teach study. I teach teachers to be teachers so that after three generations the words will not leave their hearts. The Rabbis have queried what it means to be great "in Torah." And in the pursuit of responses, the Rabbis have sought for their proof texts in the Prophets, specifically in the writings of Isaiah and Jeremiah. Obviously, for the Rabbis the Prophets are the exemplars of social leaders and teachers. The prophets remind us all, in Reb Heschel's words, that "few are guilty, but all are responsible." In this present story, Rabbi Yehuda, in the name of Rav—apparently this is not the first time someone expressed a similar puzzlement—addresses the subject of the Rabbis recent discourse: the place of suffering in the moral scheme and the centrality of study and scholarship to the moral order. Here the text cited—Jeremiah 9:11—bemoans the destruction of the land and pleads for an explanation for the desolation.

I am glad that the Rabbis continue to explore the place of suffering in human life—to try to make some sense of this terrible condition. Rav Yehudah, in the name of Rav, offers insight into these matters by elaborating the meaning of a particularly elusive and perplexing verse in Jeremiah (9:11). The prophet, Jeremiah, has been lamenting the sufferings and punishments (these recall the torments of Elazar and Rabbi of the earlier stories), which the Israelites have endured as a result of and in the experience of their exile. Jeremiah has been witness to God's declaration of anger towards the people and God's punishment of them. And the Prophet asks in puzzlement, "Who is the man that may understand this, and who is he to whom the mouth of the Lord has spoken, that he may declare it, for what the land perishes." That is, Jeremiah wonders, where is the voice that could possibly explain what has caused such total devastation of the land and people. What, the Prophet cries, could all of this mean?

It seems to me to be a straightforward question: what could have led to this suffering? Rav Yehudah said in the name of Rav that though the wise, the Sages, have discussed this issue, they could not explain it and they were unable to resolve why the land was destroyed. Nor, says the Talmud, could the Prophet Jeremiah explain why all of this has happened. But, said Rav Yehudah in the name of Rav, the answer may actually be discovered in the next verse (Jeremiah 9:12) and proffered by none other than God. Israel, God says, lies in ruins "Because of their forsaking My Torah that I put before them." Again, it would appear a simple straightforward response: the people of Israel forsook the ways of God and therefore, God punished them. That is, the failure to study—and therefore practice Torah—has led to the present sufferings. Failure to study leads to precipitous social decline and woes. In very behaviorist terms, the cause and effect are eminently visible and clear. I am not comfortable with this simplistic view; it is not consistent with the Rabbis' methods.

So it interests me again to discover that the Rabbis have offered for proof text only the first half of God's answer to Jeremiah. Theirs is a telling omission. The verse reads in full: "Because of their forsaking My Torah that I put before them, *they did not heed my voice nor follow it.*" In their selective citation, the Rabbis have here then focused their response not on the people's disobedience, but on their abandonment of the text itself. The Rabbis' omission creates a distinction between betraying the text and hearkening to the voice. It would seem, otherwise, a redundancy. If the people had *forsaken* the Torah, then it should be apparent that they have also *not heeded* God's voice. But since every word in Torah is intentional, then every word requires explication. What, I wonder, is not redundant about this verse? What is the difference between forsaking Torah and not heeding the voice? Let me propose a possible response: Though we continue to imagine worlds where books are anathema *(Farenheit 451)* and where all study is abandoned for other physical pleasures *(Brave New World, The Giver)*, though too many students boast that they have completed their schooling without completing a single book, it is impossible to consider, as these novels suggest, that all people—and hence any society—would lose completely the desire for study and for reading. At least *one* citizen would yet value books. Thus, the Rabbis reason, it could not be that the Torah was *completely* abandoned and its teachings dismissed. There must have been at least a single person remaining who did not abandon the study and practice of Torah. God had once promised Abraham that if Abraham found even ten righteous men in the city of Sodom, God would have spared the entire city. Alas, in the quintessential city of sin such citizens could not be found, but amongst the Israelite people such absolute wickedness would not be the case—after all, even in Sodom, Abraham himself was holy and he did bring Lot and his family out with him. And yet, the desolation of the land and the sufferings of the people beheld by Jeremiah give testament to the absolute ubiquity of intellectual and ethical decline.

However, the Rabbis will not accept that there was no adherence to Torah and no study of it by the population. They will not accept that not a single remnant of such engagement in scholarship remains. Thus, the complete desolation of the land remains a mystery to them. And so Rav Yehudah in the name of Rav offers a remarkable explanation. Rav said that the exile and the land's desolation was not the result of the people's failure to read the Torah, nor even that it was not studied. Indeed, the present sufferings are not necessarily even the result of the people's failure to practice ethical behavior. Rather, Rav Yehudah says in the name of Rav, the land was destroyed because the blessings that precede the study of Torah and that follow after the reading were not recited. That is, study had become secular and not sacred. Study had lost its base in wonder and awe and had become mundane. Ordinary. Prosaic. Lifeless.

You see, for the Rabbis, study is not to be undertaken for its own sake; rather, it must be related to the practical—to the attainment of holiness. Neither is study to be undertaken for the achievement of specific goals and objectives. Study is not merely a practical enterprise, but a holy one. Study is an engagement with the mysteries of the world. Study is a sacred undertaking. The blessings before and after Torah readings sacralize the event. Thus, the Rabbi's omission of the second half of the statement in Jeremiah is explained: Rav Yehudah says not that the people's behavior had become evil, but that their wonder of study had been abandoned!

And what of the activity in our classrooms today? What is it that we think we do there? How do we sacralize study? Our class sessions begin formally with a bell or series of bells and end similarly. What in our behavior hallows the work we do in the classroom? What rituals raise our work out of the mundane and into the sacred? How do we suggest to our students that what we do in classrooms is holy work? This Talmudic explication suggests that it is not study itself that is to be venerated; rather, it is the attitude toward study that raises it to the sacred. That attitude acknowledges that there is far more than we will ever know and that what we consider in the classroom is merely a hint of all that exists outside it. We ought to acknowledge in our schoolrooms that our knowledge will never suffice. We ought to stand in our classrooms in awe and wonder. We need, I think, a ritual in our classrooms that sacralizes study.

✡

Rav Hama said: "What is [the meaning of] what is written: 'In the heart of the discerning, wisdom rests: but in the inward part of fools it makes itself known'?" "In the heart of the discerning, wisdom rests"—this [refers to] a scholar, the son of a scholar. "But in the inward part of fools it makes itself known"—this [refers to] a scholar, the son of an uneducated person. Ulla said: "This is what people say: '[One] coin in a bottle cries "*kish kish*"."

Rabbi Yermeyah said to Rabbi Zera: What is [the meaning] of what is written: "The small and the great are there, and the servant is free from his master?" Do we not know that the small and the great are there? Rather, whoever makes himself small about matters of Torah in this world will be made great in the world to come. And whoever makes himself as a servant about matters of Torah in this world will be made free in the world to come.

Despite the lack of connecting phrases, these stories nonetheless seem to follow thematically and logically not only each other but those stories that have preceded as well. The Rabbis have been engaged in a far-reaching discussion that began by questioning the nature of what it means to be greater "in Torah." That discussion began with the refusal of Rabbi's proposal by Elazar's wife. The discussion has now progressed into matters concerning the nature and ends of study. There is no doubt that these Rabbis have an ulterior motive: they mean to define scholarship such that the people's engagement in it ensures both the continuance of tradition and the continued movement toward holiness. *I am holy so you shall be holy.* After all, the Rabbis have replaced the priests in the changing world order, and the family table has substituted for the sacrificial altar. In the absence of the Temple and with the elimination of the sacrifice and the priestly caste that had been responsible for accomplishing it, the Rabbis held that it would be study that would make Israel itself a nation of priests. Scholarship, engaged in with the spirit as advocated by the Rabbis, would begin *tikkun olam*—the repair of the world. And so it seems to me that in their discussion of these stories, the Rabbis search deeply into their memories and vast knowledge of the texts for references to scholars, to wisdom and to learning to elaborate their definitions of each and to associate each with holiness.

It is remarkable to me how well the Rabbis have learned their texts and move from place to place within them to construct the narrative. Above they have interpreted a verse in Jeremiah to ensure that scholarship retains its basis in ethics; study, the Rabbis argue, must remain a holy activity. Scholarship engaged in for its own sake, without awe or wonder, or which is undertaken merely to enhance the reputation of the scholar, is a desecration of scholarship itself and of the texts learned. In this first story above, the Rabbis intend to ensure the democratic nature of the scholarly enterprise. Though as we have learned, engagement in and exercise of scholarship through the generations gains permanence; the adoption of scholarship by a new generation is welcomed and legitimized. Recall that the Rabbis have shown that scholarship that persists through three generations will remain forever in that lineage. Scholarship is an important inheritance. But questions about the first generation scholar must inevitably arise: how does s/he fit into the tradition and what place is s/he permitted? Hama chooses this verse in Proverbs (14:33) to ensure that the invitation to study remains an equable and open one: all are invited to participate in study. Both the son of a scholar and the son of an uneducated person are called scholars, but inevitably their scholarship rests differently within each. To the former, scholarship has been a way of life and forms of discourse have evolved and been practiced familiarly and regularly. But for the neophyte, scholarly conversation is novel and the sound of it roars loudly at the family table. And the delight in scholarship—the attractions of it that the Rabbis knew so well—sparkle in the new scholar who proceeds marvelously in the world wondering, in the words of the folksong, "how can I keep from singing?" The Rabbis do not offer a value judgment on the preference of the one scholar over the other; what distinguishes the two is not the quality of the wisdom, but the surprise that is special and un-subduable to the new scholar. I'm not sure that the Rabbis are critical of the new scholar so much as they recognize her need to make her achievements known. A scholar from a family in which scholarship is not a tradition makes a noise, as does a coin dropped into an empty bottle. It is inevitable: how could such a scholar not be recognized? I do not think the Rabbis mean to dismiss the importance of this coin or its value. They just realistically argue that within its context it certainly makes a different noise than that dropped into, say, a rather full penny jar. Perhaps it is the development of the tradition of scholarship during the Rabbinic period, a development facilitated by the Rabbis' need to reinvent Judaism on the basis of study, at the destruction of the second Temple, which leads the Rabbis to open the tradition to all.

In the next story, the Rabbis continue to address the definition of scholarship and the scholar. This time they go to the book of Job and with remarkable consistency make the verse further their claims for scholarship. Rabbi Yermeyah is curious why this particular verse—"The small and the great are there, and the servant is free from his master"—appears in Job (3:19). Isn't it obvious, Yermeyah asks, that both the great and small will get to the world to come? Why, he wonders aloud, does the text need to comment here? In other words, in this perfect text, why does this superfluity exist? Of course, in postmodern thought, all texts are imperfect and I am postmodern enough not to suggest here that the Torah is perfect. Far better scholars than I argue for the maculateness of even the text of Torah. But I appreciate the Rabbis' questions: they interrogate not only the text's meaning but the text's contexts as well. I think that these are the questions we would teach our students to ask of text: we want our students to ask the text what it means. And we would like them to go with Yermeyah a step further; to include the context of the world in which the text lives to help them understand the verse.

That is, Yermeyah understands that the Torah exists in a world that assumes the perfection of that text; but in his question, Yermeyah asserts that the perfect text must be rendered perfect by its careful readers. If the text is perfect, it is made so by its reader. The text does not in the present case state a commonplace: the significance of the text's statement must be constructed upon a certain ideological position. But, since in the world of the Rabbis the sense of the verse in Job might appear common-place, Yermeyah assumes that meaning must be found elsewhere. And since the topic of discussion concerns the scholar and scholarship, then Yermeyah suggests in his interpretation of the text that it refers to the stance of the scholar. The scholar who understands the limits of his understanding will be always great; the scholar who remains captivated by the text will know freedom. The text, within bounds, is limitless and may take the reader anywhere the reader may carry it. Yermeyah reads as would a scholar; to be captivated by the Talmud in which Yermeyah is himself known as scholar, requires acknowledgment by the reader that there is always another explanation about the text. The scholar is she who acknowledges insufficient knowledge. The text of Talmud teaches not only how to read it but how to read other texts as well.

Job suffers. Indeed, Job has become the symbol of those who suffer seemingly without cause. And Job curses his life and bemoans his birth. Job expresses a preference for the world to come over this one. For though he suffers and is brought low *here*, Job expresses assurance that in the world to come he is assured of relief. We have returned in the present discussion to the theme of suffering. Elazar and Rabbi bore their sufferings; against his, Job rebels.

Job had come to believe that his wealth and his happiness were all in his control and were evidence of his due. He is confused and angered by his present and unwarranted suffering; he is prepared to abandon all hope. Perhaps it is that Job has ceased to be open to learning. I think that here Yermeyah refers to Job because Yermeyah rejects Job's simple cause-effect thinking: I prospered and blessed life, but now that I suffer I curse my existence. I have lived an exemplary life and am not deserving of these sufferings; if I do not believe that these sufferings are the result of my sin, I reject them and would prefer to have died in the womb. Yermeyah, I think, by quoting Job, suggests that Job holds that his knowledge is absolute and that his suffering must be unjust. Job has ceased to wonder and to stand in awe! His suffering had become great, and Job's first response to his suffering is to regret his life rather than to acknowledge the insufficiency of his knowledge.

Thus, Yermeyah interprets the verse in Job to refer to the humility of the scholar: Job is prideful and desires to pass on to the world to come, but to attain the world to come, one must first be humble before the greatness of the Torah. Job insists that he is blameless: he denies that there is more to the world than he knows, but Yermeyah offers the statement that only those who bind themselves to study here will be free in the world to come. In choosing *Job* as his proof text, Yermeyah establishes a stance and an ethics relevant to the scholar. If we scholars would all acknowledge our smallness with regard to knowledge and be the servant to the mandate for holiness, then the promise of the world to come would become the reality of the world we have now.

NOTE

1. As I have suggested, there is a gender problem with this situation that we must not explain away or condone; we can only admit that we are dealing with a patriarchal structure.

CHAPTER 9

THE NINTH SAPLING

Resh Lakish was making the [burial] caves of the Rabbis. When he reached the cave of Rabbi Hiyya, it was concealed from him. He was offended [and] said: "Master of the Universe, did I not debate about Torah like him?" A [heavenly] voice went forth and said to him: "You debated about Torah like him, [but] you did not spread Torah like him."

Again, the Rabbis return to the differences between scholars and teachers. This has been a central issue to my life: how to apportion time between study and teaching. I love to be down here in the basement reading and writing material that often remains down here in this basement. I love the unclutter of this basement study. And how is this activity considered study? In the University setting in which I now work, this discrimination is one of the most divisive issues amongst faculty. The demand to publish or perish has forced the construction of curriculum vitae to reflect not the scholar as teacher but the teacher as writer. There are many who *debate* Torah like him; I wonder how many *spread* Torah like him. I ask myself: how can I be a good teacher if I do not continue to learn? If I do not study, how might I continue to learn? If I continue to study, how will I find time to meet my students? Of course, for public school teachers, this conflict is nonexistent: the workload of which teaching is only a part, is so great that scholarship is veritably impossible.

I think that the Rabbis argue here that it is not enough to be a scholar, even, like Lakish, a great one. Rather, Torah requires dissemination, and to achieve this a teacher is necessary. I think we haven't yet moved very

far from the earlier question: what does it mean to be "in Torah." There,
I think the Rabbis earlier explored one significant aspect of this question:
what is the stance a scholar takes before study and the world. Here, the
Rabbis suggest that being "in Torah" requires more than the encyclope-
dic knowledge of Torah that the Rabbis themselves luminously display in
Talmud. The Rabbis urge that "in Torah" includes the capacity to spread
Torah. I think the Rabbis are ready to address issues central to the
teacher: what does it mean to "spread Torah"? The Rabbis exalt Rabbi
Hiyya, a brilliant scholar *and* a brilliant teacher, but, I yet wonder, how
might a brilliant teacher be defined?

✡

When Rabbi Hanina and Rabbi Hiyya would quarrel, Rabbi Hanina said
to Rabbi Hiyya: "Do you quarrel with me? If, God forbid, the Torah were
to be forgotten in Israel, I would bring it back with my argumentation."
Rabbi Hiyya said to Rabbi Hanina: "Do you quarrel with me, who has
labored for the Torah that it not be forgotten in Israel? What do I do? I go
and sow flax, and I weave nets [from it], and I catch deer, and I give their
meat to orphans. And I prepare scrolls and write the Five Books [of
Torah]. And I go to a town and teach five children the Five Books [of the
Torah], and I teach six children the six orders [of the Mishnah]. And I say
to them: "Until I come back, teach each other Torah and teach each other
Mishnah. And [thus] I labor for the Torah that it is not forgotten in
Israel." This is what Rabbi said: "How great are the deeds of Hiyya!" Rabbi
Yishmael the sons of Rabbi Yose said to him: "Even [greater] than yours,
sir?" He said to him: "Yes." "Even [greater] than my father's?" He said to
him, "God forbid! Let not such a thing be [said] in Israel."

Two sages argue about their respective superiority. Each claims to be the
better scholar-teacher. Hanina argues that he must be the better teacher
because his sagacity and rhetorical brilliance would be able to recreate
Torah even if it were completely forgotten in Israel. Hanina claims that
through his own genius, he could recreate Torah. It is interesting to consider that Hanina has not realized that if Torah were totally forgotten his
ability to recreate it might be considered a meaningless exercise. Thus, I
find Hanina's statement rather interesting considering the Rabbis' dispar-
agement of sterile scholarship above. There, it will be recalled, study separated from holiness led to desolation. Hanina has also suggested that his
genius alone is sufficient to assess his capabilities, whereas the story of
Rabbi Yohanan and Resh Lakish indicated that study requires a challenging community. Hanina assumes that his ability to recreate Torah ensures
that it will be studied: he assumes as well that Torah is, as it were, an

immaculate text. Hanina believes that his ability to engage in precise and detailed analysis of Torah ought to earn him at least a reputation equal to that of the well-respected Hiyya.

Hiyya responds passionately. His merits as a teacher derive, he exclaims, not merely from his knowledge of Torah, which is at least equal to that of Hanina, but from his ability to construct an environment suffused not only with teaching, but with learning as well. Hiyya's counter acknowledges that his work as a scholar requires students. Hiyya claims that his power and reputation as a teacher derive not only from his ability to reason engaged with a text, but from his capacities to support the environment in which the teaching of text, in this case, Torah, will occur. This includes the care and preparation of those who will continue the tradition. Whereas Hanina is concerned with preserving the past in the present, Hiyya seems engaged in transmitting the past into the future by attending to the present. Furthermore, whereas Hanina's merits maintain the tradition within himself and his immediate circle—recall that scholarship that exists through three generations in a family will never leave that family—Hiyya opens the tradition democratically. Hiyya supports and teaches the children of the poor who will then teach others.

During this sabbatical year I appreciate Hiyya's agricultural endeavors. Hiyya says that he sows the flax from which he then weaves the nets with which he catches deer whose meat feeds the orphans. Rabbi Hiyya knows that before he can teach anybody anything, they must first be ready to learn; part of his obligations as a teacher is to satisfy the basic physical needs so that learning might occur unobstructed by, say, hunger. This has long been the premise of the Head Start Program in the United States; this is the rationale for free breakfast and lunch programs in the public schools for children of economically disadvantaged families. Finally, this is the reason why all of the social plans for an improved educational system in the United States must fail: a child ready to learn must come from a family where parents have wages to provide adequate resources for life; a child ready to learn must have available high-quality day care so that the parents can earn enough money to provide amply for the child's growth; every child must come from an environment that is safe and free from regular danger. In the United States today, we are not prepared to pay for these necessities; we are not concerned with these basic needs. Rather, our educational concerns are focused on measurable outcomes that result from an education that occurs, it would appear, in a complete social vacuum. As my grandmother might say, *"Vey iz mir!"*

Hiyya is not content to merely advocate for social programs; he is himself a social program! Hiyya appears to me to epitomize the prophetic tradition to which I earlier referred; in his advocacy for those who are powerless to do so, the prophet demands social reform and justice. I recall that the Torah insistently and repeatedly obligates me to care for the orphan and the widow and the worker. Hiyya's agricultural work serves the greater purpose of enabling students to learn. Hiyya's whole life is focused on his teaching and on their learning. Of course, during this sabbatical year, my fields [of flax] I am forbidden to sow . . . however, I recall, the Sabbath produce of the land shall not only be mine to eat but available to all those who live in the household.

Hiyya's efforts continue. From the skins of the deer he prepares the scrolls of the Torah. Then he takes the five scrolls to town and teaches them to five children, then he teaches the six orders of the Mishnah to six children, and then he commands them to teach each other. I am curious whether he teaches one book to each of the five children or five books to each of the five children; personally, I embrace the latter idea. In any case, I think I have found the first teacher-educator. I think also I have found an unacknowledged source of the Lancastrian method so prevalent in the practice of education during the nineteenth century. This was a method of education specifically designed for the instruction of poor children. Joseph Lancaster, a British Quaker, opened a school for poor children, but could not afford to also hire assistants. He devised a plan whereby the older students would teach the younger. In America, the brighter students were selected as monitors—as teachers. And they were each assigned a row of pupils whom they were to supervise and instruct. The teacher taught the monitors and the monitors taught the younger students. Drill became the primary method of education.

But this story of Rabbi Hiyya suggests that simple drill, or even vast knowledge, is not as valued as the ability to spread Torah by producing others who can continue the tradition. It is not their memorization (the ability to recreate Torah) that enables Hiyya's students to teach, but their ability to disseminate the principles of interpreting Torah that makes them teachers. Hiyya's own greatness as a teacher comes not only from his critical reasoning, or complex analysis of Torah, but rather from his ability to transmit Torah in such a way that his students become its teachers. In this way Torah is spread rather than merely recreated; in this manner Torah is taught and not drilled. Hiyya teaches others how to sow flax to make the nets to catch the deer to write Torah; Hiyya teaches the five books of Torah to five students and the six tractates of Mishnah to six students who them will instruct others. Though it might appear that Hiyya teaches mere memorization, the agricultural imagery suggests that he is responsible for a great deal more. Hiyya's activities prepare the total environment of study.

That is, Hiyya's own intellectual harvest feeds his students. With the harvest that is his own learning, Hiyya weaves curriculum with which he might catch the attention of his students and satisfy their intellectual hunger. The good learner, the story says, is also the good teacher.

Hiyya acknowledges that because he labors for Torah, it will not be forgotten; Hanina attributes the persistence of Torah to his own intellectual capacity. Hanina's pride is to be distinguished from Hiyya's humility. Humility, we learned above, is a characteristic of the scholar.

✡

Rabbi Zera said: "Last night Rabbi Yose the son of Rabbi Hanina appeared to me, [and] I said to him: 'Next to whom are you placed?' He said to me: 'Next to Rabbi Yohanan.' 'And Rabbi Yohanan is next to whom?' 'Next to Rabbi Yannai.' 'And Rabbi Yannai is next to whom?' 'Next to Rabbi Hanina.' 'And Rabbi Hanina is next to whom?' 'Next to Rabbi Hiyya.' 'And Rabbi Yohanan is not next to Rabbi Hiyya?' He said to me: 'In a place of fiery sparks and burning fires, who will let the smith's son enter there?'"

Who sits next to whom, I discover here, is not only a contemporary problem. I recall that while I studied assiduously and nervously in preparation for my Bar Mitzvah, my parents struggled for hours over the seating arrangements for the party. Not a few energetic disagreements ensued. The criteria for seating arrangements, I have learned in life and books, are many and varied; the idiosyncrasies and details that are entailed are too complex to fathom. Indeed, I don't intend to explore this specific issue here, though I recall that one Long Island high school at which I taught one of the major topics of conversation amongst the faculty was the order by which parking spaces ought to be assigned. There were, after all, an upper lot and a lower lot; there were near spaces and far spaces and the senior faculty demanded near spaces in the upper lot—sometimes even if they didn't own cars. Their car pools, after all, had cars.

Teachers are often very much concerned with seating arrangements. Traditionally, students are placed in particular seats according to the place of the last name in the order of the alphabet. My children today are usually arranged by the first letter of their last names, an issue made increasingly complex by the practice of name-hyphenation. When I was in school, the spelling of my last name landed me almost always in the first seats of the first row, but the girls with whom I fell regularly in love always had names farther down in the alphabet. From my forward position, I was forced to engage in subtly executed physical contortions so that I might occasionally espy the object of my desire. The problem arose so continually that the subtlety of my efforts forever eluded the one I loved and almost always attracted the notice of the teacher. "Alan, turn around." I wonder if there is a correlation between alphabetic seating and high school romances.

Of course, this alphabetical arrangement facilitates the taking of attendance: absence is not only readily noticeable but easily identified. Alphabetized students are more easily learned: also students can be known by place rather than by effort. There are, of necessity, other factors that regulate seating arrangements in the classroom: students who experience difficulty with hearing or sight are generally assigned seats in the front rows regardless of their placement in the alphabet; the front rows are reserved as well for students the teacher must carefully watch. In more liberal classrooms, where the alphabet does not determine one's place in line, where one sits might reflect the needs for recognition: students who desire to be close to the teacher grab the front row seats while those who intend to define their distance from the teacher sit defiantly in the rear.

Nevertheless, row-seating is the traditional means of designing classroom environments: rows separate and order; rows establish ultimate visibility. In the Foulcauldian sense, rows render the student powerless and subject to the teacher's constant gaze and discipline. In rows, students become always vulnerable and therefore, always either at or subject to attention. There is little place for relaxation, or safety, or privacy in a classroom set in rows. Forever facing forward, students can be seen without seeing. In rows, movement is always visible and motives eminently assignable. Sitting in rows, students necessarily direct all conversation at the teacher; sitting in rows there is no way to have any conversation other than the prescribed one. Rows suggest that learning can only be achieved by looking at the back of the head of the person in front.

And so I am interested in the Rabbis' final comment because there seems to be a discrepant arrangement that requires explanation. Yohanan is not seated next to Hiyya, and this arrangement seems to represent a slight to Yohanan's reputation. If Hiyya is the quintessential teacher, the master scholar-educator, then Yohanan's assigned seat so distant from Hiyya represents a slight to Yohanan's reputation. Where one sits in the Heavenly Academy is a matter of rank and, as we have heard, rank is determined by learning. Zera is, therefore, astounded that Yohanan, known for his wisdom, sits so far from Hiyya in the Heavenly Academy. What could be amiss in this school of schools? Is Yohanan's reputation in learning greater than its reality? In Zera's dream, Yose offers a different rationale for the placement of Rabbi far from Hanina: Yose says, "in a place of fiery sparks and burning fires, who will let the smith's son enter there." Yose's is a rather cryptic explanation, I would maintain, but I sense that it is either for Hiyya's or Yohanon's protection that Yohanan is seated far from Hiyya. Now, the story goes that Yohanan was commonly known as Yohanan bar Nappaha—Yohanan, son of the smith; this might explain the reference to the smith as an explanation for Yohanan's striking placement away from Rabbi Hiyya. And there is also, I read, some varied opinions (as I might expect) regarding the significance of the appellation. Some say that Yohanan's father was, indeed, a smith. Others explain that this reference is a cognomen to Yohanan's great beauty. A third explanation says that Yohanan was called "Nappaha" because of this very parable now under discussion. And a fourth explanation suggests that Yohanan actually comes from a town called Nappaha. I do not know that any of these explanations offers reason for his distance from Hiyya in the seating arrangement of the Heavenly Academy as Yose's comment suggests. And so I would like to offer another explanation for the comment. There is, in Talmud, precedent for such innovation.

Yohanan is identified with the smith. In the smith's environment I would assume, the skin acquires a certain toughness as a result of the inevitable scorching and burning that must occur as a result of the sparks. I wonder: do the Rabbis keep Yohanan far from Hiyya because the toughness of Yohanan's skin defends him from a sensitivity necessary to the exemplary scholar. That is, is Yohanan too tough skinned to be admired as a scholar worthy of propinquity to Hiyya? Or do the Rabbis keep Yohanan far from Hiyya because Yohanan's tough skin makes him impenetrable: is he too hard to understand? I wonder now if tough skin becomes the scholar, or does tough skin obstruct the scholar?

Are the Rabbis concerned that the focus on scholarship could lead to a paucity of human feeling and would be detrimental to the scholar's work?

I wonder of whom I am thinking down here during the celebration of my sabbatical?

✡

Rav Haviva said: Rav Haviva bar Surmaki told me: "I saw one of the Rabbis to whom Elijah would frequently appear, whose eyes were clear in the morning but in the evening they appeared as if they had been burnt by fire. I said to him: 'What is this?' And he said to me, 'For I said to Elijah: "Show me the Rabbis as they go up to the Heavenly Academy."' He said to me: 'You may gaze upon all of them, except for the carriage of Rabbi Hiyya, upon which you may not gaze.' "What are their signs?" The angels go with all of them as they ascend and descend, except for Rabbi Hiyya's carriage which ascends and descends by itself. I was unable to restrain myself [and] I gazed upon it. Two sparks of fire came and struck that man and blinded his eyes. The next day I went and prostrated myself by his cave, [and] I said: "It is your Baraitot, sir, that I study," and I was healed.

I am reading Chaim Grade's immense two-volume novel *The Yeshiva*. Now the Yeshiva is a particularly Jewish academic institution; in it students in pairs daily study Talmud and other classical texts of Rabbinic Judaism. All day students engage in debate and disputation concerning the complexities and meanings of the texts. Moshe and Tova Halbertal say that students invest more time preparing for class than their teachers, developing their own ideas and coming to their lessons eager not only to listen and learn, but to argue and be heard. The Halbertals say, "A usual class in the Yeshiva will quickly turn from a well-ordered presentation of the teacher into a lively and sometimes chaotic exchange between a few bright students and their teacher. The classroom does not function as the presentation of the truth by the all-knowing scholar imparting knowledge to the ignorant or less knowledgeable receptacles, who write down all he says uncritically."(Halbertal, 1998) Indeed, the Yeshivah is a particular learning environment in which multi-generational conversation is played out against the backdrop of the sacred texts. And since it is interpretation and meaning that is given priority, then the ideas of no one person is overvalued; it is the unique style of teaching that is paramount in these classrooms. It is talent and knowledge that is emphasized in the Yeshivah, rather than any particular idea or theory.

All of this complexity is portrayed in Grade's novel about the development and practice of this unique educational institution and of the students who learn there. We are introduced to a wonderful panoply of characters—masters and disciples, as indeed the second volume is titled. There are no saints in this wonderful book. The more we learn about a character, the more flaws are revealed. I wonder how it is possible to learn from such flawed humans? Though all scholars daily journey to the Heavenly Academy, all but Hiyya need assistance of the angels. I appreciate the teacher's imperfections and vulnerabilities. I have often required the assistance of the angels getting to the academy.

Only Hiyya, the teacher of the year, requires no assistance. He may ascend and descend without the help of the angels. Why, however, may Hiyya not be seen in this ascent and descent? I am reminded of the story of Apollo's son who asked of his father only that he be allowed to drive the chariot of the sun. After all, it was only a horse and buggy! I think that there is an enormous difference between watching someone accomplish a task and effecting that task oneself. To someone with great skill, the task is customary. But to someone with little facility, only the ease of effort is observed and not the long process of preparation and learning. The myth of Icarus comes to mind as well: even soaring with our own father's wings has its dangers. Perhaps this Talmudic story means to caution us to remain attentive to our own works and not to the ascent (or descent) of others.

The fact is that to stare at the teacher of the year and ignore his teachings is to acknowledge the man over his work. To idolize the man is to be blinded to the work. Staring at the man in his ascent is to be blinded. Finally, it is Hiyya's Baraitot that are important. It is these teachings that can humble and heal.

How many academic processions have I observed! Being in the presence of great scholars is often exhilarating. But it is also distracting and discomfiting. I grow envious and argumentative. I lose sight of my purpose. But it is study that heals. In the midst of study, I am relieved of some of the insecurity that derives from my pedigree and lack thereof. I would keep my eyes to this world and not to the ascent and descent of the scholars.

When the great maggid, Dov Baer of Mezritch, realized that he had become known to all of the world, he begged God to tell him what sin he had committed that had brought such guilt upon him. The great scholar has great burdens, as we have understood about Rabbi Hiyya. It is only to increase the difficulty of the work of the scholar if his public posture is forever on display. The scholar, too, must be allowed her privacy and peccadilloes. It is to frustrate the scholar's purpose to exalt the scholar and not the work.

I have known many teachers in my life. What burden have I imposed on them? I have gazed at many teachers in my life. I have been struck by two sparks of fire and been blinded. I returned to their Baraitot and was healed.

✡

Elijah would frequently appear at Rabbi's academy. One day it was the first day of the month, [and] he was delayed and did not come. He said to him: What is the reason that you, sir, were delayed?" He said to him: "[I had to wait] until I awakened Abraham and washed his hands, and he prayed, and I laid him down, and similarly with Isaac, and similarly with Jacob." "But you should have awakened them together!" "I thought [that] they would pray fervently, and would bring the Messiah before his time."

I hate to wait, but I want now to think about waiting. I have always engaged in waiting; perhaps even given it serious consideration at times.

Sometimes I wonder if much of my life isn't spent in the attempt to avoid the consciousness of the inevitability of my waiting. I carry things to do—both for myself and my children—everywhere I go. I think about techniques for filling up my life in order to eschew having ever to be caught waiting. I am not alone in this concern: people all about tell me that they have little time to pause; *they* are not waiting: they are "busy." Sometimes I wonder what they would be doing if they were not busy. Wouldn't they be waiting?

Somehow now, waiting seems more an obvious presence in my life than perhaps in earlier years. I have two young children and I seem to do a lot of waiting—waiting for them to choose their clothes of a morning—this might take upwards of forty-five minutes. Waiting for them to come out of the bathtub. Twenty minutes. Waiting for them to fall asleep. Sometimes hours. Of course, waiting is endemic to Jewish experience as well; many wait eternally for the Messiah to come—and I am not certain that my attraction here to the topic of waiting is not in some clear way connected to this heritage. But I have memories as well of being introduced in 1965 while I was yet a senior in high school—by Mr. Matienzo, as I have said—to Beckett's *Waiting for Godot*. I was fascinated by these two characters dressed in Chaplinesque black who waited by a barren tree in an unidentified location for a person they had never seen who might . . . well, that is the question, isn't it? Why were they waiting? What did they expect? I saw a production of Godot that year and it has remained a vivid memory; I was steeped in the absurd my last year of high school— it prepared me perhaps, for my later life in academia, I think now. Waiting seemed central to my life: I waited to graduate and I waited to be admitted to college. I waited to be secure enough to ask Renée Lerner to be my prom date, and I waited for her to kiss me. And I waited to grow up. Back then, I liked to think that Godot was God and that these two men waited inevitably, but rather fruitlessly, for God's arrival. That was during the death of God movement that seemed to play so smoothly into the matters of the play. I pondered my fate as I waited with Vladimir and Estragon for some sign of arrival.

Today, I often feel like Vladimir and Estragon; waiting is so intrinsic to my experience. But what is to be done while I wait? *Godot* suggested perhaps, that our lives are a waiting in which time is filled. Filled with what? And how? My life's experience of waiting allows me to consider that *Waiting for Godot*, like Talmud, is about the minutiae of daily living. It is all waiting, after all, and that waiting must be filled with doing. What else is there to do? But then . . . Vladimir and Estragon never *do* anything: Estragon says, "Let's go," but the stage directions indicate:

(They do not move). Perhaps that is the discriminating issue: there seems no moral compass in the play for what there is actually to do, whereas the Talmud is constantly examining and insisting what must be done—and often in situations as absurd as the one in which Vladimir and Estragon discover themselves. But Vladimir and Estragon never actually act.

Consider the story above regarding Elijah, or the earlier stories of Elazar lying dead in the loft, yet pronouncing legal judgments. In Beckett's drama, however, since Godot will not come today, but promises to arrive tomorrow, then Vladimir and Estragon finally never have to do anything! In *Waiting for Godot* (1954), Vladimir and Estragon wait to see what it is they might do, though, indeed, Estragon suspects that there is "nothing to be done." While they wait, Vladimir and Estragon talk, and since they do nothing but wait, they can only talk about their condition of waiting. In Talmud, the Rabbis are insistent that there is a great deal to be done, and they discourse almost interminably about what it must be.

In *Waiting for Godot* (Beckett 1954, 12b), the following exchange takes place:

Vladimir: Well? What do we do?
Estragon: Don't let's do anything. It's safer.
Vladimir: Let's wait and see what he says.
Estragon: Who?
Vladimir: Godot.
Estragon: Good idea.
Vladimir: Let's wait till we know exactly how we stand.
Estragon: On the other hand it might be better to strike the iron before it freezes.
Vladimir: I'm curious to hear what he has to offer. Then we'll take it or leave it.
Estragon: What exactly did we ask him for?
Vladimir: Were you not there?
Estragon: I can't have been listening.
Vladimir: Oh . . . Nothing very definite.
Estragon: A kind of prayer.
Vladimir: Precisely
Estragon: A vague supplication
Vladimir: Exactly.
Estragon: And what did he reply?
Vladimir: That he'd see.
Estragon: That he couldn't promise anything.
Vladimir: That he'd have to think it over.
Estragon: In the quiet of his home.
Vladimir: Consult his family.

Estragon: His friends.
Vladimir: His agents.
Estragon: His correspondents.
Vladimir: His books.
Estragon: His bank account.
Vladimir: Before taking a decision.
Estragon: It's the normal thing.
Vladimir: Is it not?
Estragon: I think it is.
Vladimir: I think so too.
          Silence
Estragon: (*Anxious*) And we?
Vladimir: I beg your pardon?
Estragon: I said, And we?
Vladimir: I don't understand.
Estragon: Where do we come in?
Vladimir: Come in?
Estragon: Take your time.
Vladimir: Come in? On our hands and knees.

Now the universe in which this exchange takes place, the universe that Vladimir and Estragon understand themselves to be a part of, is uncertain, even dangerous, and certainly contingent. The dependency that Vladimir and Estragon experience, which ties them to Godot, is intrinsic to the uncertainty of what they might do in and with their lives. If there is, as they acknowledge, nothing to be done, then what is the purpose of living—what could anyone meaningfully do (or do meaningfully!) during the time before death except to bring death closer. But waiting for Godot gives the lives of Vladimir and Estragon purpose. Vladimir asserts: "What are we doing here, that is the question. And we are blessed in this, that we happen to know the answer. Yes, in this immense confusion one thing alone is clear. We are waiting for Godot to come" (51a). It is from this act of waiting that all their comfort derives. Vladimir's "We are waiting for Godot," is always answered by Estragon's sigh of relief, "Ah!" But the catch— remember, there's always a catch—while Vladimir and Estragon wait, and with nothing else to be done, they engage in thought and in talk, and in the discourse the sense of their condition overwhelms them.

With no other purpose than to wait for Godot, with no grounding in a spiritual assuredness, the two despair of their condition. "Suppose we repented," asks Vladimir. "Suppose we repented our being born." It is not that there is no spiritual sense here—indeed, the spiritual overwhelms. But this spirituality is empty: there is nothing to be done. Waiting for

Godot these two do nothing else. "All I know is that the hours are long, under these conditions, and constrain us to beguile them with proceedings which—how shall I say—which may at first sight seem reasonable, until they become a habit. You may say it is to prevent our reason from foundering. No doubt. But has it not long been straying in the night without end of the abyssal depths?" Thought is a temporary relief, but leads finally to despair. The intellectual exercises in which they engage while waiting—a waiting that comprises their lives—explore and expose the spiritual bleakness in which Vladimir and Estragon discover themselves. "And where were we yesterday evening according to you?" Vladimir asks Estragon, to which the latter responds, "How would I know? In another compartment. There's no lack of void" (42b).

Now, waiting is central to the Talmudic story told above about Elijah. Perhaps Beckett had indeed addressed an issue that has long interested humanity forever and that the Rabbis here addressed almost eighteen hundred years ago. What is life about? What should we do before we die and then can no longer do? What should we do while we are waiting to die? I am not relieved by Beckett's play. I believe, however, that Talmud offers a different perspective on the same issue—one that offers me strength and comfort and purpose.

The universe in which this Talmudic story takes place is filled with meaning. First, I note that it is set on the morning of the new moon—a time abundant with the promise of hope. Its characters are Abraham, Isaac, and Jacob, the patriarchs of Judaism about each of whom there swirls a rich body of story. The account of their lives comprise the founding ways of being in Judaism: these characters are portrayed as concrete and yet distinct men, whose responses to the world are narrated precisely and individually. Abraham courageously leaves his parents' home to settle a land far away from all that is familiar to him; Abraham digs his wells afresh. Isaac must have been devastated by his father's willingness to cast out his first-born son Ishmael and irrevocably traumatized by Abraham's near-sacrifice of his son for the father's belief. Isaac is identified to me by a certain quietism and withdrawal. He is content to be, well, detached and to devote his time to re-digging his father's wells. And finally, the third patriarch, Jacob, is portrayed as self-absorbed and conniving, a wanderer who is deceived by his uncle Laban even as he, Jacob, deceived his brother, Esau. Jacob is betrayed by his sons even as Jacob had betrayed his own father, Isaac. Jacob, I believe, is condemned to a difficult life filled with betrayal and sorrow and homeless wandering. Taken all together, the Biblical text portrays these patriarchs as nothing but fallible human beings linked, it would seem, not so much by strong family ties as by their belief in a single God and by a choice of deeds to which that belief leads them. That is, the Abraham, Isaac, and Jacob alluded to in the Talmud text exist in a world alive with meaning, with story and legend. Their histories permeate the brief Talmudic narrative. Their names and characters are no mystery.

Now Elijah too is familiar; historically, Elijah is the prophet from Gilead; he was the outstanding religious leader of his time. The Bible records that Elijah did not die, but was carried to heaven in a chariot pulled by horses of fire. Among the many stories told about the prophet Elijah, perhaps the most important identifies him as the forerunner of the Messiah. In this tradition, Elijah is charged with devising the coming time aright. This Talmudic story rests clearly in that Messianic tradition. I will return to this matter shortly. But it intrigues me that this Elijah is portrayed first as a common servant and second, as a regular visitor to the Rabbi's studio. Could it be that even Elijah continues to study? And is his study, perhaps, connected to the coming of the Messiah? Indeed, perhaps the centrality of study in Judaism, its placement at the center of faith, argues for this position.

The context in which this very brief anecdote is placed also intrigues me; this obviously apocryphal story occurs, as you may remember, in the context of a long and detailed discussion that was begun by a matter that concerned the necessary treatment of laborers by an employer. The issues there concerned the latter's obligations to the former. What does awaiting the Messiah have to do with the exigencies of labor relations? Perhaps that is, in part, the point: the Messiah is always dependent on human action, or rather, dependent on the quality of human relations. In this case, specifically, the Mishnah had stated that when engaging laborers for hire, one must always engage them and provide for them based in local custom. For example, the Mishnah continued, when Rabbi Yohanan ben Matya asked his son to contract for laborers, the son did so without exactly stipulating of what their lunch meal would consist. The father despaired: "My son, even if you make [a meal] for them like Solomon's feast in his time you have not fulfilled your obligations towards them, for they are the children of Abraham, Isaac, and Jacob." That is, in the absence of specific contractual terms based in local custom, the workers must be treated based in the whole tradition that begins with the foundational fathers, Abraham, Isaac, and Jacob.

Now, this "incident" of Rabbi ben Matya and his son is raised because in Talmud every incident teaches a precedent; this is so because it is assumed that no Sage would act unless the Halakhah on which his action was based was entirely clear and not dependent only on logical reasoning.

What could possibly trouble Rabbi ben Matya in this instance? It is the work of the Talmudic doctors to understand how Rabbi's concern is governed by the halakhic, the legal, principle. That is, ben Matya's action must be consistent with the law—halakhah; engagement in the intellectual work of understanding behavior in light of halakhah is based in clear rhetorical principles, but always, reverence for the law organizes the system of argument. And the laws are forms of behavior derived in action, they are based in the idea that life must be lived according to principle—I am holy so you should be holy—and adherence to that principle will finally make conditions right for the Messiah's coming. The halakhah, then, are practical. I appreciate, then, the idea that a story about the relations between employer and workers is somehow connected to a story about the coming of the Messiah. Perhaps it is that the Messiah cannot come until workers and laborers are at peace. As I write, I am relieved to consider that workers and management at Northwest Airlines have much to do with bringing about the Messiah.

A second aspect of this story that intrigues me, fascinates equally: Elijah explains that he has taken the more inefficient route in his actions because the conjoint praying of Abraham, Isaac, and Jacob would have been so fervent that they would have brought the Messiah before his time. Now, the Messianic theme speaks of the coming of redemption and the advent of unlimited peace and happiness. The Messianic era portends the end of exile, of suffering, and of political and economic strife. Certainly, in the time of the coming of the Messiah, labor unrest ought to be nonexistent. The Messiah offers the redemption for which the Jew waits. Not unlike Beckett's Vladimir and Estragon, who wait for Godot, the Jew waits in great expectation and hope for the Messiah. Yet oddly enough, in this story, and unlike the desires of Vladimir and Estragon, Elijah is fearful of bringing the Messiah before his time—he would rather, as it were, keep Godot off stage and thereby continue to await his arrival. In this Talmudic text, waiting is a consummation devoutly to be wished.

It is often written that Elijah can assume many appearances to accomplish his purposes; but here, however, Elijah appears as a common citizen familiar to Rabbi Yehudah ha-Nasi and as a regular visitor to his academy. It intrigues me that Elijah would be so regular a visitor here; again I wonder, could it be that the Talmud suggests that study is intrinsic even in the lives of the prophets and that even the great prophet Elijah must continue his study at Rabbi's academy? Elijah is, we are told, a daily visitor. Second, it is interesting that in this story Elijah is portrayed as servant to Abraham, Isaac, and Jacob, the traditional founding male ancestors of Judaism. He awakens them each and every day and cares for their diurnal needs. The patriarchs then pray each day—it interests me that Elijah does not appear to pray with them; rather, he is wholly preoccupied with the exigencies of their daily lives. Could it be that the Talmud suggests that the great prophet Elijah who will usher in the messianic era must live a quotidian life until he is called upon to announce the coming of the Messiah? As there is a story that the Messiah exists even now amongst us, so it is clear that Elijah, too, must exist now amongst us. It is s/he who ministers to our daily lives. Who makes it possible for us to pray, or in more contemporary terms, as it were, to study? Elijah comes to our studios—to our offices—daily. I wonder what might it be like if we treated every visitor to our offices as if they were the prophet Elijah? As if they had just come from ministering to Abraham, Isaac, or Jacob—or Sarah, Rebecca, Rachel, or Leah—performing necessary duties executed separately for each to ensure that their conjoint praying not bring the Messiah before his time? Indeed, what if we considered ourselves to be Elijah so mindful of our responsibility to others that we put off our own prayers and studies until we had completed our ministrations to the Other? I suspect that were we to think of each other and ourselves as this Elijah, the time for the coming of the Messiah would, indeed, be here.

And why not awaken the patriarchs at once? Certainly it would prove to be the more efficient method. And, it might be argued, wouldn't now be an appropriate time for the Messiah's coming? But Elijah says that the conjoined prayers of the patriarchs would bring the Messiah before s/he is due! The coming of the Messiah for whom we wait is here contingent on human action. That is, in this story the Messiah will not usher in the era of peace but will arrive at its moment. According to Talmud, the Messiah, paradoxically it would seem, must wait for us. Of course, we might say that Godot waits for Vladimir and Estragon, but they do not see it this way. And the play itself twice disappoints Vladimir and Estragon when the boy announces that Godot will not come today. "Let's go," Estragon says. (*They do not move*).

There are two possibilities concerning this story of Elijah. The first is that the coming of the Messiah will not *bring* redemption but rather, *acknowledge* it. That is, human action must prepare the world for the Messiah. The second possibility is equally as astonishing: that the Messianic era is only possible by human action. Were the patriarchs to pray in unison the Messiah might arrive, but the world would be as yet unprepared for the Messiah's coming. There is, it would seem, very much something to be done.

And so this waiting that I live with might be reconsidered as a result of this Talmudic tale. Perhaps this waiting that often seems, as it does for

Vladimir and Estragon, so interminable, is not, in fact, a waiting at all. It is, perhaps, a doing. Perchance it is not I who, in fact, waits; rather, it is the Messiah who must wait. I must act.

✡

> He [Rabbi Yehudah ha-Nasi] said to him [Elijah]: "And are there people like them in this world?" He said to him, "There are Rabbi Hiyya and his sons." Rabbi proclaimed a fast, [and] caused Rabbi Hiyya and his sons to pray. He said: "He causes the wind to blow" and a wind blew. He said "he causes the rain to fall," and the rain fell. When he was about to say: "He resurrects the dead," the world trembled. They said in Heaven, "Who revealed [our] secrets to the world?" They said: "Elijah." They brought in Elijah, [and] struck him with sixty lashes of fire. He came [and] appeared to them as a fiery bear, entered among them and drove them away.

Rabbi has asked a leading question: I think that he, at least, is ready for the coming of the Messiah. For him, it is, perhaps, time. He wonders: is there no one in this world whose conjoint praying would bring the Messiah? Well, I suppose many of us have desired that advent at various times in our lives when salvation seemed most necessary. Whatever our beliefs regarding the nature of the messianic age, which we anticipate, we desire its coming. From what, I wonder, do we each individually require saving? If wishes could bring the Messiah, I wonder now what all those wishes in aggregate might have in common that the Messiah would feel compelled to approach at such demand? Rabbi knows the compulsion the right prayer might impose on the Messiah and therefore, resorts to subterfuge to discover Elijah's secret. It is not his own prayers that would be so efficacious, but Rabbi muses, perhaps there are some . . . In the story above, in which Elijah tells of his daily attendance and ministrations to Abraham, Isaac, and Jacob, Rabbi has received a hint of the reality of salvation, and its promise seems tied to the fervency of prayer. The prayers of the patriarchs in unison would bring the Messiah; are there no others in the world whose prayers could work such wonders? Elijah knew enough not to allow the three patriarchs to pray in unison, but he discloses to Rabbi that there are such people as Abraham, Isaac, and Jacob whose prayers might bring the Messiah before his time: Rabbi Hiyya, our teacher of the year, and his sons.

Now, Rabbi Yehuda ha-Nasi's question to Elijah acknowledges Rabbi's own feelings of inadequacy. Despite his high position in the Rabbinic hierarchy, Rabbi does not believe himself one of those people whose prayers could bring the Messiah before his time. I respect Rabbi's humility. Nor, I suppose, does Rabbi maintain any hope that Elijah will ever consider bringing the patriarchs together for communal prayer to hasten the advent of the Messiah. As a later Rabbi would remark, "You don't need a weather man to know which way the wind blows." But, as a result of his position in the hierarchy, ha-Nasi does have the prerogative to command prayer and may, therefore, convene the necessary parties in the community to bring the Messiah before her time. I am comforted that in this tale the Rabbis suggest that the necessary and compelling prayers are collective and not individual: Rabbi has asked if there "are people like them in this world" whose communal prayer will effect the long-awaited advent. The attainment of the holiness to which prayer aspires requires engagement with the community. Ethics is a stance in the world occupied by and with others. Judaism seems to require community. It is said that God's presence is assured when ten are in prayer (in Judaism a quorum for prayer, a *minyan*, is ten adults), when three judge Jewish law, and when two study Torah. The Rabbis hold that certain activities require community: individual undertaking is necessary but not sufficient. I have always appreciated Judah Ha-Levi's rationale for the necessity of a *minyan*: he says that ten are required for prayer so that if one forgets the words another is there to remember. I suspect that the need of three for the judgment of the law ensures that a decision will be rendered. And two are required for study so that argument is ensured. As the story of Lakish and Yohanan above suggests, true study requires argument and disputation. I am reassured here again that the messianic era will occur not as a result of one person's actions, but from the performance of a community.

Clearly, Hiyya and his sons no longer dwell together or their regular daily prayer would have brought the Messiah long ago—albeit, I suspect, with a little reluctance. We recall, too, that scholarship maintained through three generations guarantees its perpetuity in the family. Hiyya's sons are out in the world and are doing their father's business: they are teachers, they are scholars and they are teacher-educators. Indeed, we have already established that Hiyya would have certainly earned the title teacher of the year. It is Hiyya who ministers to the physical, the intellectual, and the spiritual well-being of the student. It is the whole student, in contemporary parlance, with whom Hiyya is concerned. It is the master teacher, Hiyya, who cares for the physical well-being of those who have none other to care for them. It is Hiyya, the master-teacher, who prepares them and their materials for study. It is the master-teacher, Hiyya, who maintains the Torah not by his own solitary, albeit brilliant, discernment and rationality, but by his capacity to transmit not only the text but the text's tradition—its context—as well. It is Hiyya, the master-teacher, who knows that the maintenance of study requires the development of teachers other than himself: he teaches his students how to be themselves teachers. In so doing, Hiyya multiplies his progeny and ensures the continuance of study. The teacher of the year, Rabbi Hiyya, is a social activist, a farmer, a weaver, a hunter, a scribe, an interpreter and disseminator of text, an educational psychologist, and curriculum theorist. The teacher of the year knows that the end of study is more study; the object of education is not Torah but the study of Torah, not the text itself but the activity with text. After all, I recall, the Rabbis have decreed that the truth of Torah is not in heaven but must be discerned on earth. It is this, I suspect, that Hiyya taught his sons. I picture them sowing their own fields of flax to weave the nets to catch the deer that will feed the orphans. I see Hiyya's sons bent over their tables working on the skins of those deer and writing the texts of the books that they will teach to the children so that these children will teach others the books. The sons of this great teacher must be great teachers. And this story suggests that Hiyya's abilities to teach and learn extend to his ability to pray. It seems that the great teacher is also a great pray-er. And it is in the activity of prayer that the great teacher along with his sons might bring the Messiah before his time.

The great teacher has a great spirituality. What does it mean to be a great pray-er? I have been struggling myself with this question of late. I have always found prayer very difficult. What do I do when I pray? If prayer is only appeal, then it is directly answered with either fulfillment or disappointment. I can be disheartened that the answer to my supplications may be no and yet take comfort that at least some response is forthcoming. But perhaps prayer conceived in this manner of query-response is equivalent to common methods of evaluation: we offer what we believe our most sincere answer, and we either get the A or we don't. Alternatively, if prayer is conceptualized as praise, then what stops prayer from being mere flattery? Such worship, perhaps, finally smacks of ulterior motive.

Ah, but I know these responses are banalities. To speak with one's Deity is an incredibly complex event; it is even more so when, as Maimonides suggests, the attributes of that Deity are not human attributes.(Maimonides 1963) How to speak to a God that does not hear; how to hear a God that does not speak. And yet prayer—as study—must be more than simply talking to one's self.

But then, I have always found teaching study difficult as well. Not aversive, mind you, but difficult. I have stood in thousands of classrooms during my twenty-five years in the profession. I have gazed upon the faces and been gazed upon by the faces of thousands of students. I stand several times daily with my hand on the doorknob to a classroom, and several times daily I experience terror. Standing at the entranceway, I sense my breathing unnaturally quicken and my heart beat rather wildly. Like the narrator in Poe's "The Tell-Tale Heart," I attempt to hide my terror behind a facade of rationalism; I push open the door to my classroom and reveal my tell-tale heart. I stand before a class and am unnerved by their beating hearts. I am afraid that all of my knowledge is inadequate to the task that abides in the classroom. I am afraid that I have nothing to teach these students. I am afraid that I am alone in this classroom. Truly, I wonder, what exactly is my task as I stand in front of this classroom and how will I judge the accomplishments of the students and of myself from my position at the front. I know there are some in the standards movement for whom these questions do not exist: they seem to know exactly what teachers are supposed to teach and what students are supposed to learn. I envy their certainty though I doubt their claim. I know to what they refer, and, like Bartleby, I respond, "I know where I am"; like Bartleby I must respond, "I prefer not to." Teaching must be more than simply talking to myself in a crowd. Study is that time when I understand how much I still don't know.

In this story of Hiyya and Elijah, the Talmud suggests to me a connection between great teaching and great praying, for Hiyya is both a great teacher and a great pray-er. As any text is endlessly interpretable, then its meaning is forever "in progress" and only partially known. There is always more than the present meaning. This is what Hiyya and his sons have learned and what they must now teach, for the preparation of teachers requires not the transmission of answers but the generation of questions. Were it the opposite, then Hanina's ability to recreate Torah by himself would have earned him the teacher of the year laurel. It is here that I may seek for the relationship between praying and teaching. As it is the whole of life that comprises the practice of the teaching of Hiyya and his sons, so it must be the whole of life that comprises the activity of their prayers. And what are the contents of the prayers of Hiyya and his sons that their fervency would have brought the Messiah? Clearly the meaning of their prayers must be greater than the words, for others must have a similar quantity of them available as well. Prayer is more than the utterance; prayer must be a stance in the world from which one utters prayer. Prayer is not the answer but the question. Study, too, is not the answer but the question.

Abraham Joshua Heschel tells us that the paths to God begin in wonder. Wonder is a radical amazement; wonder is a state of maladjustment to words and notions, the recognition of their fluidity. The prayers of Rabbi Hiyya and his sons must be filled with wonder. Wonder arises in the awareness of the world's glory. The prayers of Rabbi Hiyya and his sons give utterance to the acknowledgment that all of their actions and all of their efforts only begin to approach the standard of holiness that comprises their directive: I am holy and so should you be holy. In their prayers Hiyya and his sons take a stance in awe and humility. The beginning of awe, Heschel tells us, is wonder. Prayer reminds us that there is another question that we might yet ask; prayer reminds us that there is something we have yet to comprehend; prayer reminds us that we live in a world of miracles. The beginning of wisdom is awe. Awe is the product of our wonder. And awe is the realization that our lives take place under wide horizons that span beyond the individual, the nation, the generation, and the era.

I think teaching is akin to praying, and I find teaching as difficult as praying —perhaps for similar reasons. If it is my purpose to communicate knowledge, then I might simply discourse and then evaluate whether it has been achieved. If we study only to know, then I think we take our stance in a world we believe to be finite. We either know or we do not know—evaluation will evaluate and record our knowledge. There is little opportunity and less room in these assessment measures for awe and wonder. And if we teach to simply recreate the canonical text, then it is instruction we offer and not learning. Finally, if we are learning merely to flatter, we are thinking of ourselves, and our immediate rewards, and not of learning. What we hear in response to such prayer and study is only our echo.

Rather, praying and teaching occurs on the slippery curve of the question mark; what I know serves to remind me of how much there is to learn. When I speak in the classroom, it must be with great humility. If I were to stand in the classroom with such humility, I wonder if anyone would listen?

Praying and teaching are the public personal acknowledgment of wonder and awe. Prayer and study are our response to wonder; to pray and to study is to stand in awe. Prayer and study set standards to which we might aspire but never reach. It is not for lack of trying that the standards cannot be achieved; it is that the standards always elude us. They are a consummation devoutly to be wished. I think that this is what postmodernism has been suggesting for years! It amazes me that prayer and study may be postmodern stances. For if postmodernism suggests there is no single meta-narrative—even no meta-narrative at all—then we might never know what ends our deeds may have. We stand in awe at the complexities of our lives. If chaos theory argues that there is finally order but that it is only recognizable in time, then prayer and study acknowledges our patience and our hope. Prayer and study emanate from the silence of awe and wonder. Again, I turn to Heschel: "It is not by the rare act of greatness that character is determined, but by everyday actions, by a constant effort to rend our callousness. It is constancy that sanctifies . . . Through the constant rhythm of prayers, disciplines, reminders, joys, [man] is taught not to forfeit his grandeur" (384). Prayer and study are stances to be taken in the world that derive from humility, from desire, and from hope. To pray is to know that the Messiah will come, but only when I am ready. To await the Messiah is to await the world in which justice is pursued. It must be I who pursues justice. To await the Messiah is to prepare the world for the Messiah's arrival. To await the Messiah is to know that I am meant to be better. Prayer and study is the reminder of human potential. No wonder the praying of the teacher of the year and his family might bring the Messiah before his time.

I think that it is not necessary to bring prayer into the classroom to sacralize the work we do there; study itself is the prayer.

Hanina's spirituality is to be contrasted with Hiyya's; Hanina claims to require no help to recreate a lost Torah and Hiyya can never cease working to impart it. Of what does Hanina's awe and wonder consist? Hiyya understands the necessity of the world that is required to continue study, but Hanina understands only the text outside of the context of that world—he can recreate it by his pre-Cartesian self. When Hanina prays, I wonder what are its contents?

Now, it interests me that Rabbi Yehudah ha-Nasi would arbitrarily decide that the time for the coming of the Messiah has arrived. When ha-Nasi controls the secret, he then controls the coming of the messianic era! It would be at his will that the Messiah may be brought. He has the power. And so Rabbi Yehuda ha-Nasi proclaims a fast so that Hiyya and his sons will pray together. It is time, commands ha-Nasi, for the Messiah. It interests me that the Talmud portrays Rabbi as so furtive. If Rabbi had asked Hiyya and his sons to pray together, would they not have done so? But Hiyya and his sons must have been aware that the coming of the Messiah is a process rather than a product. They must have known what Elijah above knew: that the Messiah waits for us. Rabbi would have the Messiah's advent even before the world would know what to do with the Messiah. And Rabbi knows that the teacher of the year would never assume the responsibility of bringing the Messiah; rather, the teacher of the year prepares the world that the Messiah might come. That explains Rabbi's subterfuge: had Rabbi asked them directly, Hiyya and his sons would have refused. The great teacher refuses the answer because it is the practice of study that answers.

What is the secret that Elijah reveals for which he must be punished? I think Elijah's secret consists of the idea that there is an easy means to the messianic era, and that its achievement might be accomplished by the strengths of others. It is someone else's efforts, someone else's character, and someone else's stance that will bring the Messiah; Elijah's secret is that we may remain passive and let someone else's prayers matter.

This has powerful import for today's classrooms, where knowledge is awaited and not sought and where too many students ask too few questions and offer too few responses to too few questions. They sit quietly in their seats and they respond only for our approval, or they sit silently afraid to risk any reprimand for a wrong answer. They take notes in silence. I think we must begin to teach study as the making of joyful noise.

And this Talmudic story offers us a radical alternative to the tradition of Western culture organized by the Promethean mythos. The Promethean myth is one held very dear in Western culture; the god's gift to humankind of stolen fire enraged the gods who had jealously guarded this secret from mortals. Fire, the gods knew, would give humans the resources to be godlike. Fire has long been understood as the image of knowledge—the means of light. Indeed, since the Enlightenment, the myth of Prometheus bespoke the human faith in the powers of knowledge to improve human life. The Enlightenment sparked the world with Prometheus's fire. But the unfortunate celestial benefactor Prometheus, was punished by being bound on the mountain crest where a bird of prey would daily to eat his liver. Each night that liver would be regenerated; Prometheus suffers eternally for giving fire to humanity.

But in this story of Elijah and Rabbi, knowledge *as* knowledge is not portrayed as dangerous, nor must it be secretly guarded or conceived even as particular to the gods. Indeed, as we have seen, for the Rabbis, knowledge is exactly an especially human endeavor. We recall that Maimonides argued that eating of the fruit of the Tree of Knowledge did not give humans reason; the claim that humans were made in God's image refers to the original powers of apprehension. Knowledge, the Rabbis argue, must be actively pursued and passed from generation to generation. For the Rabbis, knowledge is what we fervently seek; certainly God has not selfishly withheld it, but rather, has actually created humans beings with the capacity for it—apprehension. Knowledge is what leads humankind to its rightful end—holiness. Knowledge need not be stolen; it is a capacity and not an object.

Elijah is punished—ironically, with fire—for suggesting that there is an easy route to the Messianic era, that someone else's piety and someone else's efforts could ensure the progress and ultimate perfection of the world. Elijah's knowledge is dangerous—he advocates an abnegation of responsibility; Elijah's knowledge suggests that we need not work ourselves for the messianic era, but that we might leave the effort to others. The inflicting of sixty lashes—the number that corresponds to the sixty tractates of the Talmud—argues to me that though the prayers of Hiyya and his sons might bring the Messiah before his time, the answer to Rabbi's question forbids such easy walk to redemption. The Messiah may *not* come before his time. In this Talmudic story, knowledge is not jealously guarded; indeed, it is part of creation and freely offered, but knowledge is so only when it is undertaken in the same spirit of prayer and with great effort. Knowledge should lead to responsibility and not away from it. And so at the institution of the fast, when Hiyya and his son begin to innocently pray, it must be Elijah who must stop the praying by disrupting it with wild bears. We must all take responsibility for our actions. The Messiah cannot be brought before his time—the Messiah must wait until we are all acting as if the Messiah should be here. The classroom, I think, is where that preparation begins. The classroom is where the community expresses its wonder and stands in awe.

✡

Shmuel Yarhina'ah was Rabbi's doctor. Rabbi became ill in his eyes. He said to him: "I will place a drug [in your eyes]." He said to him: "I cannot [bear it]." [He said to him:] "I will spread an ointment [on your eyes]." He said to him: "I cannot [bear it]." He placed a vial of medicine under his pillow, and he was cured. Rabbi was anxious to ordain him, but the matter was not accomplished. He said to him, "Do not be distressed, sir. I myself have seen the Book of Adam, in which it is written: Shmuel Yarhina'ah shall be called 'scholar,' but he shall not be called 'Rabbi.' And Rabbi's recovery will be at his hand. Rabbi and Rabbi Natan are the end of the Mishnah. Rav Ashi and Ravina are the end of teaching. And your sign is: Until I came into the sanctuary of God: then I understood their end."

Rabbi, I think, has no great tolerance for pain. I like the humanity with which the Talmud portrays these Rabbis. Though historically we know them as scholarly and saintly, they are often depicted as imperfect, sometimes petty, sometimes weak—but always questing and willing to investigate further. I am glad that they are so human: I would not learn from saints.

One of my favorite passages in Thoreau's *Walden* is the opening of the chapter "The Pond in Winter." Thoreau writes:

> After a still winter night I awoke with the impression that some question had been put to me, which I had been endeavoring in vain to answer in my sleep, as what–how–when–where? But there was dawning nature, in whom all creatures live, looking in at my broad windows with serene and satisfied face, and no question on *her* lips. I awoke to an answered question, to Nature and daylight. The snow lying deep on the earth dotted with young pines, and the very slope of the hill on which my house is placed, seemed to say, Forward! Nature puts no questions and answers none which we mortals ask. She has long ago taken her resolution.(Thoreau 1970)

I often ascribe great meaning to my dreams. The twentieth-century scholars of mind after Freud have attributed to dreams, well, almost supernatural powers. Our dreams, we are told, explain our past and predict our futures; our dreams give voice to our mute unconsciousness; our dreams hold our fragile egos from disintegration. Our dreams define us and guide us: Prospero says that humans are such stuff as dreams are made of; Hamlet is afraid to dream; Raskolnikov cannot bear his dreams. In *Finnegan's Wake*, Joyce tried to rationalize a dream, depicting the world in the world of dreams.

My dreams, I think, are my production: they arise from the uninhibited release from the need to be rational. A dream may be considered an expression of self by the self's evocation through the work of the dream. In our dreams we nominate persons, objects, and events as psychically significant; we invest these objects with meaning, and that meaning evokes us as selves. And the dream work uses the objects of the world with which the waking self has established and learned relations. Dreams are where we are free to release control and become "unintegrated." This "unintegrated" state is a position of least resistance because the censorship of rationality is greatly weakened, and the dreamer may evoke a self that is relatively unrestrained. We use objects in our dreams in remarkably original ways we do not consider in our waking life and we express relationships in dreams in ways different from how they are experienced in our daily lives in the world. D. W. Winnicott argues that the dream state permits the self to be relatively free from external impingement or internal constraint. Aboriginals believe that to approach dreamtime is to become intimately attached to the original earth.

Thoreau's dreams are not prescient and answer no questions. The great issues of existence, such as where, what, how, and why may not be answered in sleep; rather, Thoreau says, it is only in the state of wide-awakedness that pursuit of such answers may occur. Perhaps that is why Thoreau next describes his morning work as cutting his way through a foot of snow to discover his drinking water. It is only by activity that sustenance is to be found. I think this is a rabbinic position. Which is not to say that Thoreau discounts his dreams. On the contrary, Thoreau acknowledges that during his stay at Walden, he learned that if one moves in the direction of one's dreams he will achieve a success unexpected in waking hours. If we have built our castles in the air, it is yet time to put the foundation under them. Our dreams lead us to waking life that is never our dream but our lives. The self evoked in dreams may be, however, then used in our waking life.

In this Talmudic story, sleep—when dreams may come—is a very powerful healer. Rabbi can tolerate neither medicine nor ointment. The pain he must endure he believes too much for him to bear: it is a pain-less restorative—sleep—which he desires. The ointment is placed underneath his pillow and while he sleeps Rabbi is healed. It is perhaps the thought of cure that relieves Rabbi's eye condition. Macbeth called sleep, "The death of each day's life, sore labour's breath, / Balm of hurt minds, great nature's second course, Chief nourisher in life's feast." Sleep is an enforced rest: it shuts out the world external and quickens the complex world within. In our dreams, the self is evoked in its engagement with objects, events, selves, and persons elected *because* they have been psychically charged within the environment during waking life. In our dreams, the self might envision itself transformed and remade using the objects with which it has related in the waking world. It is no wonder Macbeth cannot sleep: Macbeth, of course, had murdered sleep and can receive none of its benefits. No wonder Thoreau acknowledges that our lives could enact our dreams if we would only act.

Sleep is more than physically restorative; it is psychically therapeutic as well. In our sleep we are all made well. Our dreams are not mysterious nor are they magical; they are performative. They are insightful and therefore, directive and expressive. Sleep does repair and cure. I easily recall times when I have struggled desperately to comprehend what I studied only to be frustrated in that effort; in exhaustion, I fell asleep only to awaken to some resolution. Overnight the answer seemed so simply resolved and I awoke to an answered question. Often, I have at night gone to bed with some large unsettled issues yet to be considered and awakened in the morning to discover them resolved. Or at least put into some larger perspective. In the day, the action that is possible is so much less confusing than the passivity and hence, timidity of night. After all, wasn't it in the early part of this Talmud that darkness was the time when the "beasts of the forest creep forth?" Wasn't it at night that the wicked would gather themselves together in Gehinnom? The first prayer upon arising in the morning expresses gratefulness for the soul's return. Perhaps this is what the prayer means: the body animated by the soul becomes a human being. Alone, neither the body nor soul is a human being.

I wonder if in sleep the answers were achieved, or whether the sleep put the issues in perspective. Somewhere in Shakespeare (*Romeo and Juliet*, I think), sleep is called death's counterfeit, but I do not think that this is an accurate image of sleep. Mythology, modern psychology, and Talmudic stories suggest other images for sleep: there is great activity in sleep—in this rest when dreams may come.

Might I not understand this sabbatical as a sleep? I have been away from the daily world of the University. That one has been for me a life filled with pressures: I do not think a job exists that is not replete with pressure. The balloon inflates as a result of pressure; so does the bread. It is important to sometimes enjoy the luxury of unleavened bread to appreciate how beautiful the simple life can be; it is important to enjoy the luxury of unleavened bread to appreciate how rich is the texture and taste of risen dough. I have not been inside my office for almost eight months; I read only every eight pieces of mail that my dear friend Al Cheatham delivers to me. I have been asleep from the academic world, and I have become a dreaming self and the self that dreamed. I have lived that dream. I have created a self that exists by using objects that could not have been available without the sabbatical rest. I have lived in texts that are distant from those in which I have been trained and from those with which I train. I have dreamed different dreams. I have different dreams.

I will awaken from this sabbatical rest in the fall and return to the life of the University; when I awaken in the fall, that soul will return to my body and I will be thankful that soul has returned to me. Thoreau notes that he left Walden because he had other lives to live. I, too, will arise from this sabbatical to undertake new lives and create for myself new souls. When I awaken from this sabbatical in the fall I will again sow my fields and prune my vineyards and gather my crop. When I awaken from this sabbatical in the fall I will look about the world and I will know that I am healed. But during this seventh year, I will have observed a complete rest and dreamed wonderful dreams.

Of course, I must note that in this very brief story the Rabbis certify their own authority and set the limits to the canon. Oddly enough, however, the rabbis delimit their authority by distinguishing the rank of "rabbi" from that of the scholar. Rabbi wants to reward his doctor with stature; he would name him "rabbi." For whatever reason—his inability to convene or to influence the licensing board—Rabbi is not able to reward his doctor with the title. And the doctor consoles Rabbi by explaining that he has seen in the *Book of Adam*, a non-canonical text, that Shmuel Yarhina'ah shall not be called "Rabbi," but will be referred to as a scholar. His name already refers to one area of his expertise—he is an astronomer. Ironically, Shmuel's reputation has nothing to do with Rabbi's power or influence: it is by Shmuel's action that his reputation arises. Scholarship, unlike rank, must be earned and cannot be conferred. Unlike in present day academia, there are no honorary rabbinical conferrals. There is, apparently, a ranked division of labor addressed here: "rabbi" appears to be a term of political authority and leadership, and "scholar" is a descriptive term that refers to powerful knowledge. It is perhaps what Talmud has suggested all along: wisdom comes in active study and not by the issuance of degrees. And finally, there is a democratic principle at work here: unlike the monarch who can grant rank unilaterally and without justification, Rabbi has not this singular power.

Finally this story offers me comfort that certain elements of study have not changed dramatically from the era of the Amoraim—the scholars who passed on the oral tradition. The final phrases of the story represent a mnemonic device to remind students that Rav Ashi and Rab Ravina brought the amoraic period of rabbinic study to a close. Ashi and Ravina redacted the Talmud that I have been studying. The creation of this mnemonic device, as we have come to expect, comes from the Holy Books, this time from Psalms. Psalm 73 speaks of the seeming injustice of the world's order when the wicked prosper and remain unpunished. This confusion remains "Until I came into the sanctuary of God; then I understood their end." In prayer the world's wonder is recalled and the pray-er, unlike the wicked and the despicable, stands again in wonder and awe. I suspect that Psalm 73 is a good place to go for a mnemonic device: it would be a comfort to have in mind that at least theoretically, justice will prevail. In this case, we are remembering the great scholars: in Hebrew, the word for sanctuary *(mikdashei)* rhymes with Ashi, and the word for "understood," *(avinah)* rhymes with Ravina. I think my mnemonic devices were simpler: **Roy G. Biv**, for the colors of the rainbow and **Sohcoatoa** for the trigonometric relations of the angles of an isosceles triangle. Students at the St. Stanislaus School offer the following mnemonic devices to recall the place numbers to the right of the decimal point of π the number of letters in each word refers to the identity of the integer in Pi. "I hope I dance beautiful in eleven years for mommy = 3.1415926535. "I want a mouse notifying me eleven times for three weekends = 3.14159265358. On the other hand, I do not recommend this device for learning or practicing grammar.

✡

Rav Kahana said Rav Hama, the son of the daughter of Hasa, related to me: "Rabbah bar Nahmani died as a result of persecution. They informed against him in the house of the king. They said; There is a man among the Jews who causes twelve thousand Jewish men not to pay the royal poll tax a month in the summer and a month in the winter. They sent an officer of the king after him, but he did not find him. He fled and went from Pumbedita to Akra, from Akra to Agman, and from Agam to Shihin, and from Hihhin to Tzerifa, and from Tzerifa to Ena Demayin, and from Ena Defayim to Pumbedita. In Pumbedita he found him. The officer of the king happened to come to the very inn where Rabbah [was hiding]. They placed a tray before him, and they gave him two glasses to drink, and removed the tray from before him, [whereupon] his face was turned backward. They said to him [Rabbah]: 'What shall we do with him? He is the king's man!' He said to them: 'Place the tray before him, and give him one glass to drink, and remove the tray from before him, and he will recover.' They did this for him and he recovered. He said, 'I know that the man I am seeking is here.' He searched for him and he found him. He said, 'I will go from here. If they kill me, I will not reveal [where Rabbah is], but if they flog me, I will reveal it.' They brought him [Rabbah] before him, [and] he took him up to a room and bolted the door before him. He [Rabbah] prayed, the wall collapsed, he fled and went to Agma. He was sitting on the stump of a palm tree and studying."

This story and the one following tell of the death of Rabbah bar Nahmani who died as a result of persecution. In Judaism that is an old and common tale. Like the previous story, this one concerns a great teacher and provides me some insight into the values of the Talmudic doctors. The setting of this story is Pumbedita, the site of the great Rabbinic Academy already active during the time of the second Temple and established by the exiles in Babylonia. It is said that Nahmani would offer lectures in summer and winter before the pilgrimage holidays in the fall and spring. As a result of their attendance at these academic conferences, the attendees would not be at home when the tax collector would arrive. Others claim that Rabbinic scholars were exempt from paying taxes and therefore, the authorities were angry at Nahmani because he attracted so many students who then claimed exemption. I like a third explanation: it is my own. Nahmani is the third-century Henry David

Thoreau: he preached civil disobedience and would not pay his poll tax. Whereas for Thoreau the consequences was mere incarceration ("The proper place today, the only place which Massachusetts has provided for her freer and less desponding spirits, is in her prisons . . .") Nahmani is condemned to death and must flee. Thoreau was arrested walking to town; Nahmani flees from town to town to save his life.

This is an interesting story and one that calls to mind the horrifyingly comic scene in *The Exorcist* when the Devil causes Linda Blair's head to rotate. The scene also reminds me of Alice in Wonderland and the dramatic physical disturbances caused by her eating and drinking. Needless to say, and so of course I will say it, I am reminded as well of the Jefferson Airplane's song "White Rabbit" in which "one pill makes you larger and one pill makes you small." Who could have known that Western culture had drawn so seminally on Talmud! Finally, I am reminded of Elazar's earlier method for catching thieves in the tavern. I think that according to Elazar this officer might be considered a thief: He is in the tavern at the fourth hour, and he is neither a student nor a laborer. Indeed, he is out to steal Nahmani's time and rob him of his study.

Nahmani escapes through the tight network of the radical underground, protected, the narrative suggests, in safe houses of movement sympathizers. I am reminded of the dangerous journeys of runaway slaves on the Underground Railway, of the brave partisans living in the forests and making war against the Nazis during World War II; I remember the chronicled experiences in the underground of the members of Weatherman and of the well-documented flight of Patricia Hearst. But I learn that the oppression of Rabba bar Nahmani predates by many, many years these more recent flights. In hot pursuit of Rabba bar Nahmani, the king's officer, in sore need of food and drink, serendipitously pauses at the very inn where Nahmani now hides. The officer sat officiously at an empty table—leaning back in his chair with his legs flung out before him, intentionally obstructing the passageways about him, his right arm flung haughtily on the rear of the chair, his hat perhaps still atop his head, and he called loudly and rudely for strong spirits. Served his drink, he does not even grunt acknowledgment nor reach for payment. He swallows the liquid in one gulp, slams the glass on the table and orders a second. In similar fashion, he demands a third and then a fourth. He has had long days of difficult but fruitless pursuit, obstructed at each step, we would hope, by the tight network of Jewish partisans secreting Nahmani to safety. The officer is tired and incredibly frustrated. He is also probably quite irritable. The king's officer drinks more than he knows is good for him, and his face swivels and turns physically about. He can't see two feet in front of him! Some interpreters say that actually his face merely became contorted. I learn that in another piece of Talmud it is said that drinking an even number of glasses of any beverage makes one susceptible to the evil spirit. Knowing the officer's intent, the tavernkeepers made sure he was served an even number of glasses. He was appropriately stricken, as expected.

Regardless of the exact physical ailment, the king's officer suffers extreme agony and bellows loudly in panicked alarm and intemperate distress. At the sound of the officer's anguished and angry howls, the inn attendants become terrified that they could be held responsible for the fate of the king's officers, and they appeal to Rabba for advice: the very Rabba whom they have hidden for his own safety. Too often, I think, the oppressed are asked to comfort the oppressors. The innkeepers had meant to save Nahmani, but they are driven by their own fears to appeal to him to save them. They are afraid, after all, that the officer of the king will hold them accountable for his physical discomforts. Nahmani suggests that they give the man another glass to drink—the odd-numbered glass—and this drink restores the officer. The officer, being no fool, realizes that his good fortune must be a result of Nahmani's presence, and he becomes heedful of his prey's whereabouts. Though Maimonides argues that the greatest charity is that given anonymously, the Talmud here suggests that certain good deeds—as Nahmani's for the officer—might be always known.

I do not think the Rabbis would attribute to any of their fraternity the ability to do magic, which is, of course, what the story suggests is the cause of the officer's ordeal. Perhaps the Talmud means to suggest that the attendants, in the effort to protect Nahmani, have abused the officer through food and drink; they have, as it were, made him drunk. Or perhaps they have placed some poison in his food that tied his stomach up in painful knots. Nahmani, however, refuses to be saved at the expense of someone else's self-respect, and so he orders that strong coffee, or some other medicinal liquid, be brought to restore the afflicted officer and permit him to return to his duties.

No sooner is the officer cured than he realizes that his salvation was a result of the character of the very man whom he has been sent to arrest. I think the officer became aware of the danger he might have suffered at the inn and realizes that his safety was the result of Nahmani's presence. I think the officer is overwhelmed by Nahmani's selfless act. He refuses to arrest Nahmani, but acknowledges that even his own execution will not cause him to reveal Nahmani's location, though a good flogging will not prevent him from disclosing Nahmani's hiding place. Apparently, physical suffering is too much for this officer to bear; Rabbi, too, we recall, had no great tolerance for suffering. He could not bear his earlier sufferings without extraordinary bellowing, and he could not tolerate the eye ointments prescribed by his doctor. One cannot help but compare the lack of moral and physical strength of the king's officer with that of Elazar who nightly invited sufferings to his bed. One also cannot help but take note of Nahmani's flight from execution and the officer's apparently cavalier lack of fear of it. I like to think that Nahmani realizes that death ends study.

But in this situation and despite his obvious sympathy for Nahmani, the officer must make some show of duty and locks his prisoner in his room. Nahmani prays. Like the trumpets at Jericho, Nahmani's prayers bring down the walls. Then, Nahmani can return to study. Finally, it is Nahmani who must free himself, and again it is prayer that inspires his release. And when he rises from prayer, he discovers he is at study. I imagine the master Nahmani sitting on the stump of the tree surrounded by his students.

There is a wonderful story in the Haggadah. Five rabbis were gathered at Benei Berak on the night of the first Seder. All night the Rabbis, Eliezer, Yehoshua, Elazar ben Azaria, Rabbi Akiva, and Rabbi Tarfon, recounted the story of the exodus from Mitzrayim—the narrow place. Finally, towards morning, their students came to them and said, "Rabbis, the time has come for saying the morning Shema." It is said that the Rabbis were so inspired by the story of the going-out from Egypt that that night they planned the rebellion against the Roman tyranny that would result in the Bar Kochba revolt. It is said that the students warned the Rabbis of the approach of the Roman soldiers by the code phrase concerning the morning prayer. Perhaps there was no time or inclination to offer the soldiers an even number of glasses. Good teachers, like Nahmani, talk, but they must also act. Good teachers, perhaps like Nahmani, must be great rebels but not willing martyrs.

✡

They disagreed in the Heavenly Academy: If the bright spot precedes the white hair, he is ritually impure, and if the white hair precedes the bright spot, he is pure. If there is a doubt, the Holy One, blessed be He, says He is pure, and the entire Heavenly Academy says He is impure. And they said: "Who will decide [it]? Rabbah bar Nahmani will decide [it], for Rabba bar Nahmani said: 'I am alone [as an authority] in [the laws of leprosy], I am alone [as an authority] in [the laws of] tents.'" They sent a messenger after him, [but] the Angel of Death was unable to approach him, because his mouth did not cease from study. Meanwhile a wind blew and howled between the branches. He thought it was a troop of horsemen, [and] said, "Let me die, and not be delivered into the hand of the government." As he was dying he said "Pure, pure." A heavenly voice went forth and said: "Happy are you, Rabbah bar Nahmani, that your body is pure and your soul departed with [the word] pure." A note fell from Heaven in Pumbedita: "Rabba bar Nahmani has been summoned to the Heavenly Academy. Abbaye and Rava and all the Rabbis went out to attend to him, [but] they did not know where he was. They sent to Agma, [where] they saw birds hovering and casting shadows. They said: "Conclude from this [that] he is there." They eulogized him [for] three days and three nights. A note fell: "Whoever leaves will be under a ban." They eulogized him for seven days. A note fell: "Go to your homes in peace." On the day that he died, a storm lifted an Arab who was riding a camel and carried him from one side of the Pappa River and cast him to the other side: He said: "What is this?" They said to him: "Rabbah bar Nahmani has died." He said before Him: "Master of the Universe, the entire world is Yours, and Rabbah bar Nahmani is yours. Why do you destroy the world?" The storm subsided.

I have long suspected that it is at the moment of death that all questions are answered. Of course, by then it is too late. There is not even the surety that Nahmani's last word authorized the ritual purity of the afflicted person: the Torah is, after all, not in Heaven and human matters must be dealt with by human beings. Indeed, despite Nahmani's own confidence in his own knowledge, there is no evidence that his word is even accepted as the final judgment, though Nahmani's judgment regarding the purity of the other does come to be attributed—perhaps appropriately—to the state of his own soul. It is a right belief, I think, that if the questions are all answered only at death, then our puzzlements and wonderments comprise the substance of our lives. It is the regular pursuit of answers with the awareness of their un-attainability that maintains our engagement in life and sustains our work in the classroom. In the belief in the question, perhaps, all our souls might be kept pure.

Knowledge is never complete except at death. There is always more to learn. I know many things, but I have a few questions. The Rabbis went out to attend to Nahmani's burial and funeral rites. For three nights they eulogized Nahmani and for seven days they mourned his death, despite the purity of his body and soul, and despite the voice from heaven ascribing happiness to Nahmani. In Judaism, mourning is organized by the numbers: the first three days are often referred to as the days of weeping. The pain of loss is severe. Following the initial agonizing three days, a note informs them that, in fact, Nahmani must be mourned for seven days—

the period of shiva all mourners must observe upon the death of a close family member. I think these extra days assuage the extremities of loss and comfort the heart. In the process of grief we are returned to the world.

The scholar is always our close family member. Her learning, passed through three generations, ensures its maintenance forever.

✡

> Rabbi Shimon ben Halafta was a fat man. One day he was hot, [so] he went up and sat on the peak of a mountain. He said to his daughter: "My daughter, fan me with a fan and I will give you bundles of nard." In the meantime a wind blew, [and] he said: "How many bundles of nard [must I give] the Master of this [wind]?"

I recall an earlier story of fat Rabbis: there the girth of the Rabbis raised the issue of sexual potency. The Roman matron doubted the legitimacy of the Rabbi's children, claiming the size of their bellies made sexual intercourse impossible; the children of the Rabbis, she suggested, were probably bastards. Of course, by this time in the narrative, paternity has become an important issue. There, the Rabbis turned their corpulence to their advantage: big bodies mirrored the size of their sexual organs, or at least their potency. In the present story, however, body girth is not turned to the Rabbi's advantage; rather, corpulence is identified with indolence and pomposity. Indeed, I imagine Rabbi Shimon ben Halafla as a second century Sidney Greenstreet, oleaginously bargaining with Humphrey Bogart in *Casablanca* or *The Maltese Falcon*. Rabbi Shimon is not to be trusted. In this story, I see Rabbi Shimon ben Halafla flicking his wrist contemptuously at his daughter and haughtily purchasing her attention and obedience, with promise of payment for her efforts in his service with a semi-precious material essential in perfumes. I do not sense from Rabbi Shimon a great deal of respect for others.

There is, I recall, another conspicuous connection made in Jewish thought and practice with excess: on Pesach, Jews are commanded to "eat only matzos," bread that has never leavened. We know that matzah reminds us of the bread that remained unrisen at the hasty exodus from Egypt. We know that matzah reminds us of the bread of affliction that slaves ate in their bondage. But there is more to the command regarding the consumption and possession of leavened products: "For a seven-day period shall you eat matzos, but on the previous day you shall nullify the leaven from your homes; for anyone who eats leavened food—that soul shall be cut off from Israel . . . For seven days leaven may not be found in

your houses . . . You shall not eat any leavening; in all your dwellings shall
you eat matzos." Thus, the directives concerning the Pesach festival
extend beyond the symbolic association with the Hebrews during their
slavery and during the going-out; rather, the command that "no chametz
may be seen in our possession, nor may leaven be seen in your possession
in all your borders" (Exodus 13:7) suggests a far more complex relationship with leavening: a fermentation process that allows dough to rise.
Leavened bread is full of air.

Rabbi Shimon ben Halafla is full of air.

I think it is one thing not to partake of chametz and to eat only
matzah, but it is another to remove all evidence of it from within one's
boundaries and to forbid the eating of any food upon which leavening
has occurred. One explanation for this is applicable to our bloated Rabbi.
Shimon ben Halafla is fat—he has overeaten and become blown up. As
the yeast puffs up the bread, so to is Halafla puffed up with appetite, with
pride, and with lust and greed. He treats his daughter with disdain.
When we are so puffed up, we abuse ourselves and others; we seek power
over others, we forget our inclination to holiness. And so for one week
during the year, during the Pesach festival, we remove from our lives the
agent of puffiness, the yeast, and we eat only unleavened foods. Richard
Levy writes that

> The rabbis say that yeast is like the *yetzer ha-ra*, our animal inclination,
> which works insidiously inside us to thwart the aims of the *yetzer tov*, our
> divine inclination . . . We need our *yetzer ha-ra*—without our bodies,
> without appetite, without desire, we would not be human, we could not
> do any mitzvot . . . Too often, though, we don't use our *yetzer ha-ra* in the
> service of God. Therefore, one week a year, during Pesach, we take out the
> visible evidence of the *yetzer ha-ra* from our lives by cleaning our homes
> of leavened foodstuffs . . . (n.a. 1989)

During Pesach we unpuff ourselves. We reduce ourselves, as it were, by
releasing the excess air and by abstaining from that which has become
swollen. In our story, however, Rabbi Shimon ben Halafla is puffed up
and full of chametz.

You see, Shimon ben Halafla bargained scornfully with his daughter
in order to relieve his own discomfort. He believed that he has resolved
his dis-ease with a simple exchange; he has an immediate need, and he
easily quantifies its resolve. He purchases his daughter's time for his rem-
edy. He assumes mastery and control over the world with simple
exchange. And then suddenly, the wind blew, and Halalfa is stunned. He

is cooled not by some small, human act within his control, but he is soothed by some invisible force in the universe that exceeds his comprehension and remains beyond his authority. "How much nard would it require to pay the Master of the Universe for the Wind?" he wonders. Halafla is overwhelmed with his inadequacy and awed by the universe beyond his ken. He stands in wonder. Isn't this stance the starting point of prayer and study: the world is always just beyond comprehension, and we are led to wonder? We stand in awe of the universe. Every book is two that I have not yet read. Filled with answers, we are puffed up with *yetzer ha-ra*; filled with questions, we are directed by our *yetzer tov*. No longer confident in our answers we stand with humility before the world's mysteries. We are prepared to learn.

CHAPTER 10

THE TENTH SAPLING

"After I have grown old, I have had youth." Rav Hisda said, "After the flesh has become withered and the wrinkles had multiplied, the flesh was rejuvenated and the wrinkles were smoothed out, and the beauty returned to its place." It is written: "And *my lord* is old," and it is written: "And *I* am old." For the Holy One, blessed be he did not repeat what she had said. It was taught in the School of Rabbi Yishmael: "Great is peace, for even the Holy One, blessed be He, deviated [from the truth] on account of it, as it is said, 'And Sarah laughed within herself . . . "and *my lord* is old," and it is written: . . . And the Lord said to Abraham . . . "and *I* am old.""

During this sabbatical, I have come to be like Sarah. For almost thirty years, I have been a teacher. I have labored steadily in my fields, sowing and plowing and harvesting my crop. It is inevitable that they have become somewhat depleted; almost without my awareness, the produce has become slightly less rich and succulent. It is inevitable that I have grown weary. I have, perhaps, become habituated to the labor and thus, partially forgotten it. Perhaps I have lost some of my sense of wonder and delight in the fields of education and at my labors as teacher. Where once I planted my beans with enthusiasm and delight, I have since worked my fields from habit and industry. I had ceased thinking about the wonder of learning and study and begun to worry about large farms and large crops. I had become, as it were, degraded and knew the classroom as the thief of my time. The terms spent in the classroom had ironically removed me from the world of students: I ceased knowing them because

I was too busy teaching them, governed less by delight and wonder than by habit, by constraint and regulation. In the classroom, I had ceased to learn. I would sometimes enter the classrooms of my children and be taken with the delight of these young children in the schoolroom. I would return to my own exhausted fields and struggle in the depleted soils. I envied the classrooms of my children and had neither the resources nor the energies to renew my own. I had become old in the classroom. Shall I have youth again? The school year draws near. The stores begin to fill with notebooks and paper and pencils and folders and fall clothes. The penitential month of Elul approaches.

I have grown old in the classroom. Shall I have youth again? The rabbis have actually altered the text slightly to facilitate their tale: what the Rabbis present as Sarah's *statement* is in Genesis actually a *question* that expresses her incredulity that she may once again be young. The Rabbis' interpretation suggests that Sarah's statement is a reflection of the actual physical changes she experienced at the news of her impending pregnancy. If the thought of her promised pregnancy produced this effect, how much younger Sarah might look at its coming to fruition?

After I have grown old, I have had youth. I have thought wonderful thoughts during this sabbatical. I have begun to think in a new way— some of the books I had intended to read, I no longer have interest in opening. Books that I did not anticipate reading, I have studied; some of those books are even present in this work. I have lived with my children in a way I could not have anticipated; I have participated in their lives in ways I could not have anticipated. I have not had to travel far to meet so many new people, for as I changed so did they reflect my change. I had, indeed, grown old in the classroom. During this sabbatical I have had youth. I think my face is still a bit wizened and the gray continues to fleck my hair and my beard. But the wrinkles of my soul, the grooves fashioned by the waters of the years of wear have grown smooth. Of course new wrinkles have begun. "Six years you may sow your field and for six years you may prune your vineyard; and you may gather in its crop. But the seventh shall be a complete rest for the land, a Sabbath for Adonai."

Once again the Rabbis address a discrepancy in the text. And since the text is accepted as perfect though maculate, it is the Rabbi's interpretation that will reveal it so. Whether the text be considered maculate or immaculate, interpretation is requisite. The text will mean, but it requires human intervention to produce that meaning. The text contains answers but first there must be questions asked of the text. Our text offers two seemingly discrepant readings; the Rabbis must create a story that resolves the discrepancy though I do not think it is ended. Rather, it awaits another story.

When Sarah hears that she will have a child she expresses amazement because Abraham, her husband, is so old. Actually, he is one hundred years old. But when God reports to Abraham Sarah's incredulity, it is to her own aging that God refers. Actually, Sarah is ninety years old. And the Rabbis explain that God communicates this mis-information to maintain domestic family harmony. The peace of a household, *shalom bayit*, is central even to God, say the Rabbis.

I have been at home for eight months, now. In the fall I must return to the university and the classroom. I have learned during this sabbatical to live the day practicing a natural rhythm it could never have had when I was obliged to the structures of the university. The only meetings I have needed to attend were ones I called myself, and often I skipped even those. Not that for the duration of the sabbatical the myriad details that comprise this life disappeared; rather, it was possible to accomplish them in a more relaxed fashion. I have let the weeds grow as flowers. During this sabbatical I never had need of an alarm clock: whenever I awakened that is when the day began. What does awaken us in the morning? Perhaps it is the anticipation of the day. I often went to sleep awaiting some aspect of the writing day to which I aspired. I think sometimes my dreams would inspire awakeness. But perhaps the two are, after all, the same thing. If my work is not engaged in the fulfillment of the dream, then it is mere drudgery and I would not labor in those fields. If the dream does not lead me to labor, then I must never awaken.

The Sabbath produce of the land was ours to eat—for my family and myself. In the quiet of early morning, when the household was yet asleep, I descended to the writing desk and engaged in study. Footsteps above me, or the record of time at the bottom of the screen recalled me; though I am on sabbatical the rest of the world continues. My children are upstairs, and when they arise, the requirements of their day will become my duties: meals and breakfasts and schools and activities and violins and soccer and homework and innumerable games of Go Fish. It's okay. I have learned during this sabbatical that my children are not my fields and that I need never rest from their care. Indeed, as we have eaten of the Sabbath produce of the land, we have together grown stronger. They have replenished my resources.

The peace of this household was at times endangered by my presence. Perhaps one must learn to live in a house. During this sabbatical I think I have grown accustomed to my place of dwelling.

I wonder how I have changed as a result of this sabbatical. The Rabbis above have changed slightly the text of Torah; the Rabbis state that Sarah remarks on the physical changes that have already taken place, but Sarah has actually asked a question: Shall I become young again after I have been old? Her question concerns the future, the Rabbis report the present. The clichés regarding the connection between future and present are too numerous to repeat here, and they are probably all true. I think that during this sabbatical I have learned to refocus my vision. In the absence of all political pressures from the university setting, I have been free to direct my energies elsewhere. I am not younger but freer. I have eaten the Sabbath produce of the land. I and all the residents of my household.

✡

> And she said: "Who would have said to Abraham that Sarah would nurse children?" How many children did Sarah nurse? Rabbi Levi said: On that day that Abraham weaned his son Isaac, he made a great feast. All the peoples of the world were complaining and saying: "Have you seen [that] old man and old woman who brought [home] a foundling from the market and [now] say: 'He is our son'? And moreover they make a great banquet to strengthen their words." What did Abraham our father do? He went and invited all the great men of the generation, and Sarah our mother invited their wives. And each one brought her child with her, but did not bring her wet nurse. And a miracle was performed with Sarah our mother, and her breasts opened like two springs, and she nursed them all. But still they were murmuring and saying: "Granted that 'Sarah who is ninety years old gave birth,' [but] did Abraham who is one hundred years old beget a child?" Immediately, Isaac's facial features changed and he looked like Abraham. They all began and said: "Abraham begot Isaac."

What forms of skepticism will greet me when I return from my sabbatical? I will perhaps look not much changed from when I left for my sabbatical, though I have since given birth. "So," they will ask, "What did you accomplish with your sabbatical?" I do not know how I might answer this query. By all accountings I will be even older than when I had last exited the office. If they have not changed, how could they treat me any differently than when I had taken leave? Will my offspring be known as mine, or will they be considered foundlings?

Yet still doubting the legitimacy of the child, the guests whisper that even if they acquiesce to the idea that Sarah *might* have been capable of conceiving a child (and they are not prepared to so readily concede), the people yet remain convinced that it could not be by Abraham, who was even older than Sarah. Suddenly, the Rabbis state, Isaac's face changed, took on the visage of his father, and his paternity became visible to all. This, the Rabbis state, is the meaning of the seemingly redundant phrase in Genesis 25:19 that says, "Abraham begot Isaac." The Rabbis say that this refers not to the actual birthing, or even the sexual coupling in which Sarah was impregnated, but to the remarkable similarity of features. That is, Isaac looked exactly like his father.

I live in a house where no one looks like me. Perhaps they never will, but sometimes I see a resemblance. At the dinner table the other day I told the family a joke. They laughed politely, but Emma, the older child, turned to me and said, "That was narrated nicely." I saw my features in her words. Perhaps all of us look like Abraham and Sarah. There are no miracles in education. Education itself is a miracle. To learn is to acknowledge that we did not know before. Education is the movement into something new. We may not always look different but we always are so. Skepticism be damned.

A story is told of the Maggid of Zlotchov.
One of his disciples asked: "In the book of Elijah it is said that 'Everyone in Israel is duty bound to say: "When will my work approach the works of my fathers and mothers, Abraham and Sarah, Isaac and Rebecca, and Jacob and Leah and Rachel?"' "How is it possible," the disciple asked, "to ever imagine that we could achieve what our ancestors achieved?" And the Maggid responded: "Just as our fathers and mothers invented new ways of serving, each a new service according to his or her own character: one the service of love, the other that of stern justice, the third that of beauty, so each one of us in our own way shall devise something new in the light of the teachings and of service, and do what has not yet been done."(Buber 1967)

We all can look like ourselves and yet continue the work of the patriarchs and matriarchs. This form of education would not have as its objectives the achievement of standards set in the past, but rather, it would aspire to learn to act now in the service of love, in the pursuit of justice and the celebration of beauty. In this way, we always would look like our parents and we would always look like our children. We would always look like our students and our teachers. Our physical appearances would be inconsequential.

✡

Rather, before they begin work, go out and say to them: "On condition that you have no [claim] on me except for bread and beans, etc." Rav Aha the son of Rav Yosef said to Rav Hisda: "Did we learn 'bread [made] of beans,' or did we learn 'bread and beans?'" He said to him a *vav* is needed like a runner on the Librut.

We've come back to the beans! Beans, beans the musical fruit, the more you eat the more you toot.

Thoreau, too, writes about beans. During his two years at Walden Pond, he maintained a bean field. He hoed and cared for it religiously. He plowed his fields, sowed his crop and harvested his beans. I am certain that Thoreau realized great satisfaction from this work, but he came to realize that the intensity of his effort far exceeded the value of it. I think Thoreau learned that there were seeds other than beans that might be sown and whose harvest might better improve his existence, however mean that existence might be. In the effort of his labor, Thoreau realized its value to his life. In his engagement in work, Thoreau (1970, 292) re-evaluated his life. "I said to myself: I will not plant beans . . . with so much industry another summer, but such seeds, if the seed is not lost, as sincerity, truth, simplicity, faith, innocence, and the like, and see if they will not grow in this soil, even with less toil and manurance, and sustain me, for surely it has not been exhausted for these crops." Thoreau went into his fields to learn about his life.

I have been on sabbatical. I have, as it were, left my bean fields alone. When I return to them in the fall, as I must—perhaps, as I will—I would again plant seeds, but perhaps not with the same intensity of effort as in the past. Or even with the same intention. Indeed, I have during my sabbatical obtained other seeds, strange, different seeds nurtured in another time and place, and they would produce novel fruit. I would see how these new seeds might grow in the soil and how they can nourish us. I will sow them this Autumn and see if they will not blossom with even less toil and manurance to sustain us, for surely the soil has not been exhausted for these crops. I think I would rather cultivate the character of our students rather than their academic standards; I would teach them to nurture others. I teach for a world where high standards of justice and mercy are expected rather than for one whose scores on standardized tests are higher. I would my students strove for a holiness we have been commanded to achieve. I would their classrooms be sacred and that their study be prayer.

We would as teachers in schools make the world ready for the Messiah even though when the world is ready, we will already know it. During my sabbatical I have learned to renew my labor.

✡

Rabbi Shimon ben Gamliel says: "He did not need [to say this]: everything is in accordance with local custom." What does "everything" add? It adds what we have learned: If someone hired a worker and said to him: "[I will pay you] like one or two of the townspeople,' he gives him the lowest wage. [These are] the words of Rabbi Yehoshua. But the Sages say: We take an average between them.

In this final section of Talmud, I suppose there is a resolution. But I don't have it.

And how does a book end? How do I finish this? As the Talmud finishes? So inconclusively?

REFERENCES

Abram, D. 1997. *The Spell of the Sensuous*. New York: Vintage Books.

Althusser, L. 1971. *Lenin and Philosophy* (B. Brewster, trans.). New York: Monthly Review Press.

Beckett, S. 1954. *Waiting for Godot*. New York: Grove Press.

Block, A. A. 1999. "Curriculum as Affichiste." In W. F. Pinar (ed.), *New Curriculum Identities*. New York: Peter Lang.

Block, A. A. (Winter, 1999-2000). "Curriculum from the Back of the Book Store." *Encounter: Education for Social Justice* 12(4): 17–27.

Block, A. A. (1997, Fall). "Finding Lost Articles: The Return of Curriculum." *Journal of Curriculum Theorizing, 13*(3), 5–12.

Block, A. A. 2004. *Talmud, Curriculum, and the Practical: Joseph Schwab and the Rabbis*. New York: Peter Lang.

Bollas, C. 1992. *Being a Character*. New York: Hill and Wang.

Buber, M. 1947. *Tales of the Hasidim*. New York: Schocken Books.

Buber, M. 1967. *On Judaism* (N. Glatzer, ed.). New York: Schocken Books.

Doctorow, E. L. 1971. *The Book of Daniel*. New York: New American Library.

Fiedler, L. 1991. *Fiedler on the Roof*. Boston: David R. Godine, Publisher.

Giroux, H. A. 1992. *Border Crossings*. New York: Routledge.

Grade, C. 1976. *The Yeshiva*. New York: Macmillan Publishing Company, Incorporated.

Grafton, A. 1997. *The Footnote*. Cambridge, Massachusetts: Harvard University Press.

Gramsci, A. 1971. *Selections from the Prison Notebooks* (Q. Hoare and G. N. Smith, eds. and trans). New York: International Publishers.

Halbertal, M. 1997. *People of the Book*. Harvard, Massachusetts: Harvard University Press.

Halbertal, M., and Tova H. 1998. "The Yeshiva." In A. O. Rorty (ed.), *Philosophers on Education* (pp. 458–469). New York: Routledge.

Halivni, D. W. 1997. *Revelation Restored*. Boulder, Colorado: Westview Press.

Hauptman, J. 1998. *Rereading the Rabbis: A Woman's Voice*. Boulder, Colorado: Westview Press.

Heschel, A. 1955. *God in Search of Man*. New York: Farrar, Straus and Giroux.

Heschel, A. J. 1951. *The Sabbath*. New York: Farrar, Straus and Giroux.

Heschel, A. J. 1959. *Between God and Man* (F. A. Rothschild, ed.). New York: The Free Press.

Keating, T. 1999. *Saturday School*. Bloomington, Indiana: Phi Delta Kappa Educational Foundation.

Kohl, H. 1994. *"I Won't Learn From You," and Other Thoughts on Creative Maladjustment*. New York: New Press.

Kraemer, D. 1996. *Reading the Rabbis: The Talmud as Literature*. New York: Oxford University Press.

Levinas, E. 1990. *Difficult Freedoms* (S. Hand, trans.). Baltimore, MD: Johns Hopkins University Press.

Levinas, E. 1994. *Outside the Subject* (M. B. Smith, trans.). Stanford, CA: Stanford University Press.

Maimonides, M. 1963. *The Guide of the Perplexed* (S. Pines, trans.). Chicago, Illinois: University of Chicago Press.

Mann, H. 1957. *The Republic and the School* (L. Cremin, ed.). New York: Teachers College Press.

Messerli, J. 1972. *Horace Mann: A Biography*. New York: Alfred A. Knopf, Inc.

n.a. 1989. *On Wings of Freedom* (R. Levy, Ed. & Trans). Hoboken, NJ: Ktav Publishing House, Inc.

n.a. 2002. *Pirke Avot* (M. R. Davis, ed.). Brooklyn, N.Y.: Mesorah Heritage Foundation.

Nissenbaum, S. 1996. *The Battle for Christmas*. New York: Vintage Books.

Ouaknin, M. A. 1995. *The Burnt Book* (L. Brown, trans.). Princeton, New Jersey: Princeton University Press.

Pinar, W. F. 1994. *Autobiography, Politics and Sexuality: Essays in Curriculum Theory*. New York: Peter Lang.

Pinar, W. F., and Madeleine R. Grumet 1976. *Toward a Poor Curriculum*. Dubuque, Iowa: Kendall/Hunt Publishing Company.

Purpel, D. 1989. *The Moral and Spiritual Crisis in Education*. New York: Bergin & Garvey.

Ravitch, D. 1995. *National Standards in American Education: A Citizen's Guide*. Washington, D.C.: Brookings Publishing.

Smith, F. 1990. *To Think*. New York: Teachers College Press.

Steinsaltz, A. 1976. *The Essential Talmud* (C. Galai, trans.). New York: Basic Books.

Steinsaltz, A. 1992. *In the Beginning* (Y. Hanegbi, ed. and trans). Northvale, New Jersey.

Stone, I. 1989. *The Trial of Socrates*. New York: Alfred A. Knopf.

Thoreau, H. D. 1970. *The Annotated Walden* (P. Van Doren Stern, ed.). New York: Marlboro Books Corp.

Vonnegut, K. 1998. *Cat's Cradle*. New York: Henry Holt.

Vygotsky, L. 1988. *Thought and Language* (A. Kozulin, Trans.). Cambridge, Massachusetts: The MIT Press.

Waskow, A., & Berman, P. O. 1996. *Tikkun: New Jewish Stories to Heal the Wounded World*. New York: Jason Aronson Publishers.

Wieseltier, L. 1998. *Kaddish*. New York: Alfred A. Knopf, Inc.

Winnicott, D. 1971. *Playing and Reality*. New York: Routledge.

Wisconsin Department of Public Instruction. n.d. *Wisconsin Model Academic Standards*. http://dpi.wi.gov/standards/index.html (accessed August 31, 2006).

Yerulshami, Y. 1982. *Zakhot: Jewiosh History and Jewish Memory*. Seattle: University of Washington Press.

Zahorski, K. J. 1994. *The Sabbatical Mentor: A Practical Guide to Successful Sabbaticals*. Bolton, MA: Anker Publishing Company.

CPSIA information can be obtained at www.ICGtesting.com
Printed in the USA
LVOW042150280812

296422LV00004B/14/P